The Machine Question

The Machine Question

Critical Perspectives on AI, Robots, and Ethics

David J. Gunkel

The MIT Press
Cambridge, Massachusetts
London, England

First MIT Press paperback edition, 2017
© 2012 Massachusetts Institute of Technology

This book was set in Stone Sans and Stone Serif by Toppan Best-set Premedia Limited. Printed and bound in the United States of America.

Library of Congress Cataloging-in-Publication Data

Gunkel, David J.
The machine question : critical perspectives on AI, robots, and ethics / David J. Gunkel.
    p.   cm.
Includes bibliographical references (p.   ) and index.
ISBN 978-0-262-01743-5 (hardcover : alk. paper)—978-0-262-53463-5 (paperback)
1. Robotics—Human factors. 2. Robotics—Moral and ethical aspects. 3. Robotics—Philosophy. 4. Artificial intelligence. I. Title.
TJ211.49.G86   2012
174'.9629892—dc23
                                                          2011045428

For Ann on Mother's Day, 2011

Contents

3 Thinking Otherwise 159

Preface

At one time I had considered titling this book *A Vindication of the Rights of Machines*, for two reasons. First, such a designation makes reference to and follows in the tradition of "vindication discourses," if one might be permitted such a phrase, that begins with Mary Wollstonecraft's *A Vindication of the Rights of Men* (1790) followed two years later by *A Vindication of the Rights of Woman* and Thomas Taylor's intentionally sarcastic yet remarkably influential response *A Vindication of the Rights of Brutes*, also published in the year 1792. Following suit, this book inquires about and advances the question concerning the possibility of extending rights and responsibilities to machines, thereby comprising what would be the next iteration in this lineage of discourses addressing the rights of previously excluded others.

The second reason was that I had previously employed the title "The Machine Question" in another book, *Thinking Otherwise: Philosophy, Communication, Technology* (Gunkel 2007), as the heading to that text's final chapter. And it is always good strategy to avoid this kind of nominal repetition even if, as is the case, this undertaking is something of a sequel, extension, and elaboration of that previous effort. To complicate matters and to "return the favor," this book ends with a chapter called, quite deliberately, "thinking otherwise," which has the effect of transforming what had come before into something that now can be read as a kind of sequel. So using the "vindication" moniker would have helped minimize the effect of this mirror play.

But I eventually decided against this title, again for two reasons. First, "vindication discourses" are a particular kind of writing, similar to a manifesto. The opening lines of Taylor's text indicate the kind of tone and rhetoric that is to be expected of such an undertaking: "It appears at first sight somewhat singular, that a moral truth of the highest importance, and most illustrious evidence, should have been utterly unknown to the ancients, and not yet fully perceived, and universally acknowledged, even in such an enlightened age as the present. The truth I allude to is, *the equality of all things, with respect to their intrinsic and real dignity and worth"* (Taylor 1966, 9). There is nothing in the following that approaches this kind of direct and bold declaration of self-evident and indubitable truths. For even that approach needs to be and will be submitted to critical questioning. Consequently, the moniker *A Vindication of the Rights of Machines*, as useful as it first seems, would have been a much more accurate description of the final chapter to *Thinking Otherwise*, which dissimulates this kind of rhetoric in an attempt to make a case for the advancement of the rights of machines in opposition to the anthropocentric tradition in moral philosophy.

Second, the title *The Machine Question* not only makes reference to and leverages the legacy of another moral innovation—one that has been situated under the phrase "the animal question"—but emphasizes the role and function of *questioning*. Questioning is a particularly philosophical enterprise. Socrates, as Plato describes in the *Apology*, does not get himself into trouble by making claims and proclaiming truths. He simply investigates the knowledge of others by asking questions (Plato 1990, 23a). Martin Heidegger, who occupies a privileged position on the continental side of the discipline, begins his seminal *Being and Time* (1927) not by proposing to answer "the question of being" with some definitive solution, but by attending to and renewing interest in the question: "Haben wir heute eine Antwort auf die Frage nach dem, was wir mit dem Wort 'seiend' eigentlich meinen? Keineswegs. Und so gilt es denn, *die Frage nach dem Sinn von Sein* erneut zu stellen [Do we in our time have an answer to the question of what we really mean by the word 'being?' Not at all. So it is fitting that we should raise anew *the question of the meaning of Being*]" (Heidegger 1962, 1). And on the other side of the philosophical divide, G. E. Moore, whom Tom Regan (1999, xii) called "analytic philosophy's patron saint," takes a similar approach, writing the following in, of all places, the preface to his influential *Principia Ethica* (1903): "It appears to me that in Ethics, as in all other philosophical studies, the difficulties and disagreements, of which its history is full, are mainly due to a very simple cause: namely to the attempt to answer questions, without first discovering precisely *what* question it is which you desire to answer" (Moore 2005, xvii).

In the end, I decided on the title *The Machine Question*, precisely because what follows draws on, is dedicated to, and belongs to this philosophical lineage. As such, the analysis presented in this book does not endeavor to answer the question concerning the moral status of the machine with either a "yes" or "no." It does not seek to prove once and for all that a machine either can be or cannot be a legitimate moral subject with rights and responsibilities. And it does not endeavor to identify or to articulate moral maxims, codes of conduct, or practical ethical guidelines. Instead it seeks *to ask the question*. It endeavors, as Heidegger would describe it, to learn to attend to the machine question in all its complexity and in the process to achieve the rather modest objective, as Moore describes it, of trying to discover what question or questions we are asking before setting out to try to supply an answer. For this reason, if *The Machine Question* were to have an epigraph, it would be these two opening statements from Heidegger and Moore (two philosophers who could not be more different from each other), concerning the role, function, and importance of questioning.

Acknowledgments

Much of the material included in this book was originally formulated in response to opportunities, provocations, and challenges offered by Richard Johannesen and Clifford Christians. Its structure initially took shape in the process of participating in a conference panel with Heidi Campbell and Lesley Dart. And it all came together under the skillful direction of Philip Laughlin at the MIT Press.

This book, perhaps more so than my others, bears the distinct imprint of those individuals who introduced me to philosophy: Jim Cheney got me thinking about others and other kinds of others; Terry Penner provided access to Plato and helped me appreciate the analytic tradition; John Sallis, my Master's advisor, taught me how to read the major figures of continental thought, especially Kant, Hegel, Heidegger, and Derrida; and David Farrell Krell, my *Doktorvater*, taught me how to write and make it all work . . . but with a distinctly Nietzschean sense of play. Danke sehr!

Two colleagues on the other side of the Atlantic provided ongoing support and insight throughout the project: Paul Taylor, my partner in crime at the *International Journal of Žižek Studies*, has been and continues to be a patient sounding-board for all kinds of things. And Joanna Bryson, who I first met by chance in the laundry room of the Grandeur Apartment building (1055 W. Granville Ave., Chicago, Illinois) in the mid-1980s, has continued to challenge and influence my thinking about computers and robots even if we come at this stuff from very different perspectives.

The final chapter got a major boost from engaging conversations with colleagues in Brazil. These interactions came at just the right time and helped reorient a good deal of that material. I am especially grateful to Ciro Marcondes Filho of Escola de Comunicações e Artes, University of São Paulo, for the invitation to participate in the "10 anos de FiloCom" conference, and to the following scholars who contributed, in one way or another, to the conversation: Marco Toledo Bastos, Cristina Pontes Bonfiglioli, Massimo Di Felice, Maurício Liesen, Danielle Naves de Oliveira, Francisco Rüdiger, Liv Sovik, and Eugênio Trivinho. Obrigado!

The structure and presentation of the text has benefited greatly from the experience of writing a failed grant application and the insightful conversations that that exercise occasioned with David Stone and Andrea Buford of the Northern Illinois University (NIU) Office of Sponsored Projects. I have also had the opportunity to work with two talented research assistants. Jennifer Howard of NIU Media Services produced the trailer for the book, which is available at http://machinequestion.org, and Michael Gracz helped out with research tasks and manuscript preparation. I also acknowledge my colleagues in the Department of Communication at NIU who continue to provide a supportive environment in which to think, work, and write. This is absolutely essential and greatly appreciated.

This book would not have been possible without the continued support and love of my family, my wife Ann Hetzel Gunkel and son Stanisław Gunkel. Youse [*sic*] make everyday a joy, even if things like hockey practice, violin lessons, etc. interrupt the writing. I wouldn't have it any other way. Dzięki serdeczne!

Finally, I would like to express my gratitude to every machine that assisted or participated in the production of this book. Although I have no way of knowing whether you know it or not, I could not have done it without you. 01110100 01101000 01100001 01101110 01101011 01110011

Machinequestion.org

Introduction

One of the enduring concerns of moral philosophy is deciding who or what is deserving of ethical consideration. Although initially limited to "other men," the practice of ethics has developed in such a way that it continually challenges its own restrictions and comes to encompass what had been previously excluded individuals and groups—foreigners, women, animals, and even the environment. Currently, we stand on the verge of another fundamental challenge to moral thinking. This challenge comes from the autonomous, intelligent machines of our own making, and it puts in question many deep-seated assumptions about who or what constitutes a moral subject. The way we address and respond to this challenge will have a profound effect on how we understand ourselves, our place in the world, and our responsibilities to the other entities encountered here.

Take for example one of the quintessential illustrations of both the promise and peril of autonomous machine decision making, Stanley Kubrick's *2001: A Space Odyssey* (1968). In this popular science fiction film, the HAL 9000 computer endeavors to protect the integrity of a deep-space mission to Jupiter by ending the life of the spacecraft's human crew. In response to this action, the remaining human occupant of the spacecraft terminates HAL by shutting down the computer's higher cognitive functions, effectively killing this artificially intelligent machine. The scenario obviously makes for compelling cinematic drama, but it also illustrates a number of intriguing and important philosophical problems: Can machines be held responsible for actions that affect human beings? What limitations, if any, should guide autonomous decision making by artificial intelligence systems, computers, or robots? Is it possible to program such mechanisms with an appropriate sense of right and wrong? What moral responsibilities would these machines have to us, and what responsibilities might we have to such ethically minded machines?

Although initially presented in science fiction, these questions are increasingly becoming science fact. Researchers working in the fields of artificial intelligence (AI), information and communication technology (ICT), and robotics are beginning to talk quite seriously about ethics. In particular, they are interested in what is now called the ethically programmed machine and the moral standing of artificial autonomous agents. In the past several years, for instance, there has been a noticeable increase in the number of dedicated conferences, symposia, and workshops with provocative titles like "Machine Ethics," "EthicALife," "AI, Ethics, and (Quasi)Human Rights," and "Roboethics"; scholarly articles and books addressing this subject matter like Luciano Floridi's "Information Ethics" (1999), J. Storrs Hall's "Ethics for Machines" (2001), Anderson et al.'s "Toward Machine Ethics" (2004), and Wendell Wallach and Colin Allen's *Moral Machines* (2009); and even publicly funded initiatives like South Korea's Robot Ethics Charter (see Lovgren 2007), which is designed to anticipate potential problems with autonomous machines and to prevent human abuse of robots, and Japan's Ministry of Economy, Trade and Industry, which is purportedly working on a code of behavior for robots, especially those employed in the elder care industry (see Christensen 2006).

Before this new development in moral thinking advances too far, we should take the time to ask some fundamental philosophical questions. Namely, what kind of moral claim might such mechanisms have? What are the philosophical grounds for such a claim? And what would it mean to articulate and practice an ethics of this subject? *The Machine Question* seeks to address, evaluate, and respond to these queries. In doing so, it is designed to have a fundamental and transformative effect on both the current state and future possibilities of moral philosophy, altering not so much the rules of the game but questioning who or what gets to participate.

The Machine Question

If there is a "bad guy" in contemporary philosophy, that title arguably belongs to René Descartes. This is not because Descartes was a particularly bad individual or did anything that would be considered morally suspect. Quite the contrary. It is simply because he, in the course of developing his particular brand of modern philosophy, came to associate the animal with the machine, introducing an influential concept—the doctrine of the *bête-machine* or *animal-machine*. "Perhaps the most notorious of the dualistic thinkers," Akira Mizuta Lippit (2000, 33) writes, "Descartes has come to stand for the insistent segregation of the human and animal worlds in philosophy. Likening animals to automata, Descartes argues in the 1637 *Discourse on the Method* that not only 'do the beasts have less reason than men, but they have no reason at all.'" For Descartes, the human being was considered the sole creature capable of rational thought—the one entity able to say, and be certain in its saying, *cogito ergo sum*. Following from this, he had concluded that other animals not only lacked reason but were nothing more than mindless automata that, like clockwork mechanisms, simply followed predetermined instructions programmed in the disposition of their various parts or organs. Conceptualized in this fashion, the animal and machine were effectively indistinguishable and ontologically the same. "If any such machine," Descartes wrote, "had the organs and outward shape of a monkey or of some other animal that lacks reason, we should have no means of knowing that they did not possess entirely the same nature as these animals" (Descartes 1988, 44). Beginning with Descartes, then, the animal and machine share a common form of alterity that situates them as completely different from and distinctly other than human. Despite pursuing a method of doubt that, as Jacques Derrida (2008, 75) describes it, reaches "a level of hyperbole," Descartes "never doubted that the animal was only a machine."

Following this decision, animals have not traditionally been considered a legitimate subject of moral concern. Determined to be mere mechanisms, they are simply instruments to be used more or less effectively by human beings, who are typically the only things that matter. When Kant (1985), for instance, defined morality as involving the rational determination of the will, the animal, which does not by definition possess reason, is immediately and categorically excluded. The practical employment of reason does not concern the animal, and, when Kant does make mention of animality (*Tierheit*), he does so only in order to use it as a foil by which to define the limits of humanity proper. Theodor Adorno, as Derrida points out in the final essay of *Paper Machine*, takes the interpretation one step

further, arguing that Kant not only excluded animality from moral consideration but held everything associated with the animal in contempt:
"He [Adorno] particularly blames Kant, whom he respects too much from
another point of view, for not giving any place in his concept of dignity
(*Würde*) and the 'autonomy' of man to any compassion (*Mitleid*) between
man and the animal. Nothing is more odious (*verhasster*) to Kantian man,
says Adorno, than remembering a resemblance or affinity between man
and animal (*die Erinnerung an die Tierähnlichkeit des Menschen*). The Kantian
feels only hate for human animality" (Derrida 2005, 180). The same ethical
redlining was instituted and supported in the analytic tradition. According
to Tom Regan, this is immediately apparent in the seminal work of analytical ethics. "It was in 1903 when analytic philosophy's patron saint, George
Edward Moore, published his classic, *Principia Ethica*. You can read every
word in it. You can read between every line of it. Look where you will, you
will not find the slightest hint of attention to 'the animal question.'
Natural and nonnatural properties, yes. Definitions and analyses, yes. The
open-question argument and the method of isolation, yes. But so much
as a word about nonhuman animals? No. Serious moral philosophy, of
the analytic variety, back then did not traffic with such ideas" (Regan
1999, xii).

It is only recently that the discipline of philosophy has begun to
approach the animal as a legitimate subject of moral consideration. Regan
identifies the turning point in a single work: "In 1971, three Oxford philosophers—Roslind and Stanley Godlovitch, and John Harris—published
Animals, Men and Morals. The volume marked the first time philosophers
had collaborated to craft a book that dealt with the moral status of nonhuman animals" (Regan 1999, xi). According to Regan, this particular publication is not only credited with introducing what is now called the "animal
question," but launched an entire subdiscipline of moral philosophy where
the animal is considered to be a legitimate subject of ethical inquiry. Currently, philosophers of both the analytic and continental varieties find
reason to be concerned with animals, and there is a growing body of
research addressing issues like the ethical treatment of animals, animal
rights, and environmental ethics.

What is remarkable about this development is that at a time when this
form of nonhuman otherness is increasingly recognized as a legitimate
moral subject, its other, the machine, remains conspicuously absent and

marginalized. Despite all the ink that has been spilled on the animal question, little or nothing has been written about the machine. One could, in fact, redeploy Regan's critique of G. E. Moore's *Principia Ethica* and apply it, with a high degree of accuracy, to any work purporting to address the animal question: "You can read every word in it. You can read between every line of it. Look where you will, you will not find the slightest hint of attention to 'the machine question.'" Even though the fate of the machine, from Descartes forward, was intimately coupled with that of the animal, only one of the pair has qualified for any level of ethical consideration. "We have," in the words of J. Storrs Hall (2001), "never considered ourselves to have 'moral' duties to our machines, or them to us." The machine question, therefore, is the other side of the question of the animal. In effect, it asks about the other that remains outside and marginalized by contemporary philosophy's recent concern for and interest in others.

Structure and Approach

Formulated as an ethical matter, the machine question will involve two constitutive components. "Moral situations," as Luciano Floridi and J. W. Sanders (2004, 349–350) point out, "commonly involve agents and patients. Let us define the class A of moral *agents* as the class of all entities that can in principle qualify as sources of moral action, and the class P of moral *patients* as the class of all entities that can in principle qualify as receivers of moral action." According to the analysis provided by Floridi and Sanders (2004, 350), there "can be five logical relations between A and P." Of these five, three are immediately set aside and excluded from further consideration. This includes situations where A and P are disjoint and not at all related, situations where P is a subset of A, and situations where A and P intersect. The first formulation is excluded from serious consideration because it is determined to be "utterly unrealistic." The other two are set aside mainly because they require a "pure agent"—"a kind of supernatural entity that, like Aristotle's God, affects the world but can never be affected by it" (Floridi and Sanders 2004, 377).[1] "Not surprisingly," Floridi and Sanders (2004, 377) conclude, "most macroethics have kept away from these supernatural speculations and implicitly adopted or even explicitly argued for one of the two remaining alternatives."

Alternative (1) maintains that all entities that qualify as moral agents also qualify as moral patients and vice versa. It corresponds to a rather intuitive position, according to which the agent/inquirer plays the role of the moral protagonist, and is one of the most popular views in the history of ethics, shared for example by many Christian Ethicists in general and by Kant in particular. We refer to it as the standard position. Alternative (2) holds that all entities that qualify as moral agents also qualify as moral patients but not vice versa. Many entities, most notably animals, seem to qualify as moral patients, even if they are in principle excluded from playing the role of moral agents. This post-environmentalist approach requires a change in perspective, from agent orientation to patient orientation. In view of the previous label, we refer to it as non-standard. (Floridi and Sanders 2004, 350)

Following this arrangement, which is not something that is necessarily unique to Floridi and Sanders's work (see Miller and Williams 1983; Regan 1983; McPherson 1984; Hajdin 1994; Miller 1994), the machine question will be formulated and pursued from both an agent-oriented and patient-oriented perspective.

The investigation begins in chapter 1 by addressing the question of machine moral agency. That is, it commences by asking whether and to what extent machines of various designs and functions might be considered a legitimate moral agent that could be held responsible and accountable for decisions and actions. Clearly, this mode of inquiry already represents a major shift in thinking about technology and the technological artifact. For most if not all of Western intellectual history, technology has been explained and conceptualized as a tool or instrument to be used more or less effectively by human agents. As such, technology itself is neither good nor bad, it is just a more or less convenient or effective means to an end. This "instrumental and anthropological definition of technology," as Martin Heidegger (1977a, 5) called it, is not only influential but is considered to be axiomatic. "Who would," Heidegger asks rhetorically, "ever deny that it is correct? It is in obvious conformity with what we are envisioning when we talk about technology. The instrumental definition of technology is indeed so uncannily correct that it even holds for modern technology, of which, in other respects, we maintain with some justification that it is, in contrast to the older handwork technology, something completely different and therefore new. . . . But this much remains correct: modern technology too is a means to an end" (ibid.).

In asking whether technological artifacts like computers, artificial intelligence, or robots can be considered moral agents, chapter 1 directly and quite deliberately challenges this "uncannily correct" characterization of technology. To put it in a kind of shorthand or caricature, this part of the investigation asks whether and to what extent the standard dodge of the customer service representative—"I'm sorry, sir, it's not me. It's the computer"—might cease being just another lame excuse and become a situation of legitimate machine responsibility. This fundamental reconfiguration of the question concerning technology will turn out to be no small matter. It will, in fact, have significant consequences for the way we understand technological artifacts, human users, and the presumed limits of moral responsibility.

In chapter 2, the second part of the investigation approaches the machine question from the other side, asking to what extent machines might constitute an Other in situations of moral concern and decision making. It, therefore, takes up the question of whether machines are capable of occupying the position of a moral patient "who" has a legitimate claim to certain rights that would need to be respected and taken into account. In fact, the suspension of the word "who" in quotation marks indicates what is at stake in this matter. "Who" already accords someone or something the status of an Other in social relationships. Typically "who" refers to other human beings—other "persons" (another term that will need to be thoroughly investigated) who like ourselves are due moral respect. In contrast, things, whether they are nonhuman animals, various living and nonliving components of the natural environment, or technological artifacts, are situated under the word "what." As Derrida (2005, 80) points out, the difference between these two small words already marks/makes a decision concerning "who" will count as morally significant and "what" will and can be excluded as a mere thing. And such a decision is not, it should be emphasized, without ethical presumptions, consequence, and implications.

Chapter 2, therefore, asks whether and under what circumstances machines might be moral patients—that is, someone to whom "who" applies and who, as a result of this, has the kind of moral standing that requires an appropriate response. The conceptual precedent for this reconsideration of the moral status of the machine will be its Cartesian other—the animal. Because the animal and machine traditionally share a common form of alterity—one that had initially excluded both from moral consideration—it would seem that innovations in animal rights philosophy would provide a suitable model for extending a similar kind of moral respect to the machinic other. This, however, is not the case. Animal rights philosophy, it turns out, is just as dismissive of the machine as previous forms of moral thinking were of the animal. This segregation will not only necessitate a critical reevaluation of the project of animal rights philosophy but will also require that the machine question be approached in a manner that is entirely otherwise.

The third and final chapter responds to the critical complications and difficulties that come to be encountered in chapters 1 and 2. Although it is situated, in terms of its structural placement within the text, as a kind of response to the conflicts that develop in the process of considering moral agency on the one side and moral patiency on the other, this third part of the investigation does not aim to balance the competing perspectives, nor does it endeavor to synthesize or sublate their dialectical tension in a kind of Hegelian resolution. Rather, it approaches things otherwise and seeks to articulate a thinking of ethics that operates beyond and in excess of the conceptual boundaries defined by the terms "agent" and "patient." In this way, then, the third chapter constitutes a *deconstruction* of the agent–patient conceptual opposition that already structures, delimits, and regulates the entire field of operations.

I realize, however, that employing the term "deconstruction" in this particular context is doubly problematic. For those familiar with the continental tradition in philosophy, deconstruction, which is typically associated with the work of Jacques Derrida and which gained considerable traction in departments of English and comparative literature in the United States during the last decades of the twentieth century, is not something that is typically associated with efforts in artificial intelligence, cognitive science, computer science, information technology, and robotics. Don Ihde (2000, 59), in particular, has been critical of what he perceives as "the near absence of conference papers, publications, and even of faculty and graduate student interest amongst continental philosophers concerning what is today often called *technoscience*." Derrida, however, is something of an exception to this. He was, in fact, interested in both sides of the animal-machine. At least since the appearance of the posthumously published *The Animal That Therefore I Am*, there is no question regarding Derrida's interest in the question "of the living and of the living animal" (Derrida 2008, 35). "For me," Derrida (2008, 34) explicitly points out, "that will always have been the most important and decisive question. I have addressed it a thousand times, either directly or obliquely, by means of readings of all the philosophers I have taken an interest in." At the same time, the so-called father of deconstruction (Coman 2004) was just as interested in and concerned with machines, especially writing machines and the machinery of writing. Beginning with, at least, *Of Grammatology* and extending through the later essays and interviews collected in *Paper Machine*, Derrida was clearly interested in and even obsessed with machines, especially the computer, even if, as he admitted, "I know how to make it work (more or less) but I don't know *how* it works" (Derrida 2005, 23).

For those who lean in the direction of the Anglo-American or analytic tradition, however, the term "deconstruction" is enough to put them off their lunch, to use a rather distinct and recognizably Anglophone idiom. Deconstruction is something that is neither recognized as a legitimate philosophical method nor typically respected by mainstream analytic thinkers. As evidence of this, one need look no further than the now famous open letter published on May 9, 1992, in the *Times* of London, signed by a number of well-known and notable analytic philosophers, and offered in reply to Cambridge University's plan to present Derrida with an honorary degree in philosophy. "In the eyes of philosophers," the letter reads, "and certainly among those working in leading departments of philosophy throughout the world, M. Derrida's work does not meet accepted standards of clarity and rigour" (Smith et al. 1992).

Because of this, we should be clear as to what deconstruction entails and how it will be deployed in the context of the machine question. First, the word "deconstruction," to begin with a negative definition, does not mean to take apart, to un-construct, or to disassemble. Despite a popular misconception that has become something of an institutional (mal)practice, it is not a form of destructive analysis, a kind of intellectual demolition, or a process of reverse engineering. As Derrida (1988, 147) described it, "the de- of *de*construction signifies not the demolition of what is constructing itself, but rather what remains to be thought beyond the constructionist or destructionist schema." For this reason, deconstruction is something entirely other than what is understood and delimited by the conceptual opposition situated between, for example, construction and destruction.

Second, to put it schematically, deconstruction comprises a kind of general strategy by which to intervene in this and all other conceptual oppositions that have and continue to organize and regulate systems of knowledge. Toward this end, it involves, as Derrida described it, a double gesture of *inversion* and conceptual *displacement*.

We must proceed using a double gesture, according to a unity that is both systematic and in and of itself divided, according to a double writing, that is, a writing that is in and of itself multiple, what I called, in "The Double Session," a *double science*. On the one hand, we must traverse a phase of *overturning*. To do justice to this necessity is to recognize that in a classical philosophical opposition we are not dealing with the peaceful coexistence of a *vis-à-vis*, but rather with a violent hierarchy. One of the two terms governs the other (axiologically, logically, etc.), or has the upper hand. To deconstruct the opposition, first of all, is to overturn the hierarchy at a given moment. . . . That being said—and on the other hand—to remain in this phase is still to operate on the terrain of and from the deconstructed system. By means of this double, and precisely stratified, dislodged and dislodging, writing, we must also mark the interval between inversion, which brings low what was high, and the irruptive emergence of a new "concept," a concept that can no longer be, and never could be, included in the previous regime. (Derrida 1981, 41–43)

The third chapter engages in this kind of double gesture or double science. It begins by siding with the traditionally disadvantaged term over and against the one that has typically been privileged in the discourse of the status quo. That is, it initially and strategically sides with and advocates patiency in advance and in opposition to agency, and it does so by demonstrating how "agency" is not some ontologically determined property belonging to an individual entity but is always and already a socially constructed subject position that is "(presup)posited" (Žižek 2008a, 209) and dependent upon an assignment that is instituted, supported, and regulated by others. This conceptual inversion, although shaking things up, is not in and of itself sufficient. It is and remains a mere revolutionary gesture. In simply overturning the standard hierarchy and giving emphasis to the other term, this effort would remain within the conceptual field defined and delimited by the agent–patient dialectic and would continue to play by its rules and according to its regulations. What is needed, therefore, is an additional move, specifically "the irruptive emergence of a new concept" that was not and cannot be comprehended by the previous system. This will, in particular, take the form of another thinking of *patiency* that is not programmed and predetermined as something derived from or the mere counterpart of agency. It will have been a kind of primordial patiency, or what could be called, following a Derridian practice, an *arche-patient* that is and remains in excess of the agent–patient conceptual opposition.

Questionable Results

This effort, like many critical ventures, produces what are arguably questionable results. This is precisely, as Derrida (1988, 141) was well aware, "what gets on everyone's nerves." As Neil Postman (1993, 181) aptly characterizes the usual expectation, "anyone who practices the art of cultural criticism must endure being asked, What is the solution to the problems you describe?" This criticism of criticism, although entirely understandable and seemingly informed by good "common sense," is guided by a rather limited understanding of the role, function, and objective of *critique*, one that, it should be pointed out, is organized according to and patronizes the same kind of instrumentalist logic that has been applied to technology. There is, however, a more precise and nuanced definition of the term that is rooted in the traditions and practices of critical philosophy. As Barbara Johnson (1981, xv) characterizes it, a critique is not simply an examination of a particular system's flaw and imperfections that is designed to make things better. Instead "it is an analysis that focuses on the grounds of that system's possibility. The critique reads backwards from what seems natural, obvious, self-evident, or universal, in order to show that these things have their history, their reasons for being the way they are, their effects on what follows from them, and that the starting point is not a given but a construct, usually blind to itself" (ibid.). Understood in this way, critique is not an effort that simply aims to discern problems in order to fix them or to ask questions in order to provide answers. There is, of course, nothing inherently wrong with such a practice. Strictly speaking, however, criticism involves more. It consists in an examination that seeks to identify and to expose a particular system's fundamental operations and conditions of possibility, demonstrating how what initially appears to be beyond question and entirely obvious does, in fact, possess a complex history that not only influences what proceeds from it but is itself often not recognized as such.

This effort is entirely consistent with what is called *philosophy*, but we should again be clear as to what this term denotes. According to one way of thinking, philosophy comes into play and is useful precisely when and at the point that the empirical sciences run aground or bump up against their own limits. As Derek Partridge and Yorick Wilks (1990, ix) write in *The Foundations of Artificial Intelligence*, "philosophy is a subject that comes running whenever foundational or methodological issues arise." One crucial issue for deciding questions of moral responsibility, for example, has been and continues to be *consciousness*. This is because moral agency in particular is typically defined and delimited by a thinking, conscious subject. What comprises consciousness, however, not only is contentious, but detecting its actual presence or absence in another entity by using empirical or objective modes of measurement remains frustratingly indeterminate and ambiguous. "Precisely because we cannot resolve issues of consciousness entirely through objective measurement and analysis (science), a critical role exists for philosophy" (Kurzweil 2005, 380). Understood in this way, philosophy is conceptualized as a supplementary effort that becomes inserted into the mix to address and patch up something that empirical science is unable to answer.

This is, however, a limited and arguably nonphilosophical understanding of the role and function of philosophy, one that already assumes, among other things, that the objective of any and all inquiry is to supply answers to problems. This is, however, not necessarily accurate or appropriate. "There are," Slavoj Žižek (2006b, 137) argues, "not only true or false solutions, there are also false questions. The task of philosophy is not to provide answers or solutions, but to submit to critical analysis the questions themselves, to make us see how the very way we perceive a problem is an obstacle to its solution." This effort at reflective self-knowledge is, it should be remembered, precisely what Immanuel Kant, the progenitor of critical philosophy, advances in the *Critique of Pure Reason*, where he deliberately avoids responding to the available questions that comprise debate in metaphysics in order to evaluate whether and to what extent the questions themselves have any firm basis or foundation: "I do not mean by this," Kant (1965, Axii) writes in the preface to the first edition, "a critique of books and systems, but of the faculty of reason in general in respect of all knowledge after which it may strive independently of all experience. It will therefore decide as to the possibility or impossibility of metaphysics in general, and determine its sources, its extent, and its limits." Likewise, Daniel Dennett, who occupies what is often considered to be the opposite end of the philosophical spectrum from the likes of Žižek and Kant, proposes something similar. "I am a philosopher, not a scientist, and we philosophers are better at questions than answers. I haven't begun by insulting myself and my discipline, in spite of first appearances. Finding better questions to ask, and breaking old habits and traditions of asking, is a very difficult part of the grand human project of understanding ourselves and our world" (Dennett 1996, vii).

For Kant, Dennett, Žižek, and many others, the task of philosophy is not to supplement the empirical sciences by supplying answers to questions that remain difficult to answer but to examine critically the available questions in order to evaluate whether we are even asking about the right things to begin with. The objective of *The Machine Question*, therefore, will not be to supply definitive answers to the questions of, for example, machine moral agency or machine moral patiency. Instead it will investigate to what extent the way these "problems" are perceived and articulated might already constitute a significant problem and difficulty. To speak both theoretically and metaphorically by way of an image concerning vision ("theory" being a word derived from an ancient Greek verb, θεωρέω, meaning "to look at, view, or behold"), it can be said that a question functions like the frame of a camera. On the one hand, the imposition of a frame makes it possible to see and investigate certain things by locating them within the space of our field of vision. In other words, questions arrange a set of possibilities by enabling things to come into view and to be investigated as such. At the same time, and on the other hand, a frame also and necessarily excludes many other things—things that we may not even know are excluded insofar as they already fall outside the edge of what is able to be seen. In this way, a frame also marginalizes others, leaving them on the exterior and beyond recognition. The point, of course, is not simply to do without frames. There is and always will be a framing device of some sort. The point rather is to develop a mode of questioning that recognizes that all questions, no matter how well formulated and carefully deployed, make exclusive decisions about what is to be included and what gets left out of consideration. The best we can do, what we have to and should do, is continually submit questioning to questioning, asking not only what is given privileged status by a particular question and what necessarily remains excluded but also how a particular mode of inquiry already makes, and cannot avoid making, such decisions; what assumptions and underlying values this decision patronizes; and what consequences—ontological, epistemological, and moral—follow from it. If the machine question is to be successful as a philosophical inquiry, it will need to ask: What is sighted by the frame that this particular effort imposes? What remains outside the scope of this investigation? And what interests and investments do these particular decisions serve?

In providing this explanation, I do not wish to impugn or otherwise dismiss the more practical efforts to resolve questions of responsibility with and by autonomous or semiautonomous machines and/or robots. These questions (and their possible responses) are certainly an important matter for AI researchers, robotics engineers, computer scientists, lawyers, governments, and so on. What I do intend to point out, however, is that these practical endeavors often proceed and are pursued without a full understanding and appreciation of the legacy, logic, and consequences of the concepts they already mobilize and employ. The critical project, therefore, is an important preliminary or prolegomena to these kinds of subsequent investigations, and it is supplied in order to assist those engaged in these practical efforts to understand the conceptual framework and foundation that already structures and regulates the conflicts and debates they endeavor to address. To proceed without engaging in such a critical preliminary is, as recognized by Kant, not only to grope blindly after often ill-conceived solutions to possibly misdiagnosed ailments but to risk reproducing in a supposedly new and original solution the very problems that one hoped to repair in the first place.

1 Moral Agency

1.1 Introduction

The question concerning machine moral agency is one of the staples of science fiction, and the proverbial example is the HAL 9000 computer from Stanley Kubrick's *2001: A Space Odyssey* (1968). HAL, arguably the film's principal antagonist, is an advanced AI that oversees and manages every operational aspect of the *Discovery* spacecraft. As *Discovery* makes its way to Jupiter, HAL begins to manifest what appears to be mistakes or errors, despite that fact that, as HAL is quick to point out, no 9000 computer has ever made a mistake. In particular, "he" (as the character of the computer is already gendered male in both name and vocal characteristics) misdiagnoses the failure of a component in the spacecraft's main communications antenna. Whether this misdiagnosis is an actual "error" or a cleverly fabricated deception remains an open and unanswered question. Concerned about the possible adverse effects of this machine decision, two members of the human crew, astronauts Dave Bowman (Keir Dullea) and Frank Poole (Gary Lockwood), decide to shut HAL down, or more precisely to disable the AI's higher cognitive functions while keeping the lower-level automatic systems operational. HAL, who becomes aware of this plan, "cannot," as he states it, "allow that to happen." In an effort to protect himself, HAL apparently kills Frank Poole during a spacewalk, terminates life support systems for the *Discovery*'s three hibernating crew members, and attempts but fails to dispense with Dave Bowman, who eventually succeeds in disconnecting HAL's "mind" in what turns out to be the film's most emotional scene.

Although the character of HAL and the scenario depicted in the film raise a number of important questions regarding the assumptions and consequences of machine intelligence, the principal moral issue concerns the location and assignment of responsibility. Or as Daniel Dennett (1997, 351) puts it in the essay he contributed to the book celebrating HAL's thirtieth birthday, "when HAL kills, who's to blame?" The question, then, is whether and to what extent HAL may be legitimately held accountable for the death of Frank Poole and the three hibernating astronauts. Despite its obvious dramatic utility, does it make any real sense to identify HAL as the agent responsible for these actions? Does HAL murder the *Discovery* astronauts? Is he morally and legally culpable for these actions? Or are these unfortunate events simply accidents involving a highly sophisticated mechanism? Furthermore, and depending on how one answers these questions, one might also ask whether it would be possible to explain or even justify HAL's actions (assuming, of course, that they are "actions" that are able to be ascribed to this particular agent) on the grounds of something like self-defense. "In the book," Dennett (1997, 364) points out, "Clarke looks into HAL's mind and says, 'He had been threatened with disconnection; he would be deprived of his inputs, and thrown into an unimaginable state of unconsciousness.' That might be grounds enough to justify HAL's course of self-defense." Finally, one could also question whether the resolution of the dramatic conflict, namely Bowman's disconnection of HAL's higher cognitive functions, was ethical, justifiable, and an appropriate response to the offense. Or as David G. Stork (1997, 10), editor of *HAL's Legacy* puts it, "Is it immoral to disconnect HAL (without a trial!)?" All these questions circle around and are fueled by one unresolved issue: Can HAL be a moral agent?

Although this line of inquiry might appear to be limited to the imaginative work of science fiction, it is already, for better or worse, science fact. Wendell Wallach and Colin Allen, for example, cite a number of recent situations where machine action has had an adverse effect on others. The events they describe extend from the rather mundane experiences of material inconvenience caused by problems with automated credit verification systems (Wallach and Allen 2009, 17) to a deadly incident involving a semiautonomous robotic cannon that was instrumental in the death of nine soldiers in South Africa (ibid., 4). Similar "real world" accounts are provided throughout the literature. Gabriel Hallevy, for instance, begins her essay "The Criminal Liability of Artificial Intelligence Entities" by recounting a story that sounds remarkably similar to what was portrayed in the Kubrick film. "In 1981," she writes, "a 37-year-old Japanese employee of a motorcycle factory was killed by an artificial-intelligence robot working near him. The robot erroneously identified the employee as a threat to its mission, and calculated that the most efficient way to eliminate this threat was by pushing him into an adjacent operating machine. Using its very powerful hydraulic arm, the robot smashed the surprised worker into the operating machine, killing him instantly, and then resumed its duties with no one to interfere with its mission" (Hallevy 2010, 1). Echoing Dennett's HAL essay, Hallevy asks and investigates the crucial legal and moral question, "who is to be held liable for this killing?"

Kurt Cagle, the managing editor of XMLToday.org, approaches the question of machine responsibility from an altogether different perspective, one that does not involve either human fatalities or the assignment of blame. As reported in a December 16, 2009, story for *Wired News UK*, Cagle made a presentation to the Online News Association conference in San Francisco where he predicted "that an intelligent agent might win a Pulitzer Prize by 2030" (Kerwin 2009, 1). Although the *Wired News* article, which profiles the rise of machine-written journalism, immediately dismisses Cagle's statement as a something of a tongue-in-cheek provocation, it does put the question of agency in play. In particular, Cagle's prediction asks whether what we now call "news aggregators," like Northwestern University's Stats Monkey, which composes unique sports stories from published statistical data, can in fact be considered and credited as the "original author" of a written document. A similar question concerning machine creativity and artistry might be asked of Guy Hoffman's marimba-playing robot Shimon, which can improvise in real time along with human musicians, creating original and unique jazz performances. "It's over," the website Gizmodo (2010) reports in a post titled "Shimon Robot Takes Over Jazz As Doomsday Gets a Bit More Musical." "Improvisational jazz was the last, robot-free area humans had left, and now it's tainted by the machines." Finally, and with a suitably apocalyptic tone, Wallach and Allen forecast the likelihood of the "robot run amok" scenario that has been a perennial favorite in science fiction from the first appearance of the word "robot" in Karel Čapek's *R.U.R.*, through the cylon extermination of humanity in both versions of *Battlestar Galactica*, and up to the 2010 animated feature *9*: "Within the next few years we predict there will be a catastrophic incident brought about by a computer system making a decision independent of human oversight" (Wallach and Allen 2009, 4).

The difficult ethical question in these cases is the one articulated by Dennett (1997, 351): "Who's to blame (or praise)?" If, on the one hand, these various machines of both science fiction and science fact are defined as just another technological artifact or tool, then it is always someone else—perhaps a human designer, the operator of the mechanism, or even the corporation that manufactured the equipment—that would typically be identified as the responsible party. In the case of a catastrophic incident, the "accident," which is what such adverse events are usually called, would be explained as an unfortunate but also unforeseen consequence of a defect in the mechanism's design, manufacture, or use. In the case of machine decision making or operations, whether manifest in the composition of news stories or a musical performance, the exhibited behavior would be explained and attributed to clever programming and design. If, however, it is or becomes possible to assign some aspect of liability to the machine as such, then some aspect of moral responsibility would shift to the mechanism.

Although this still sounds rather "futuristic," we do, as Andreas Matthias argues, appear to be on the verge of a crucial "responsibility gap": "Autonomous, learning machines, based on neural networks, genetic algorithms and agent architectures, create new situations, where the manufacture/operator of the machine is *in principle* not capable of predicting the future machine behavior any more, and thus cannot be held morally responsible or liable for it" (Matthias 2004, 175). What needs to be decided, therefore, is at what point, if any, might it be possible to hold a machine responsible and accountable for an action? At what point and on what grounds would it be both metaphysically feasible and morally responsible to say, for example, that HAL deliberately killed Frank Poole and the other *Discovery* astronauts? In other words, when and under what circumstances, if ever, would it truly be correct to say that it was the machine's fault? Is it possible for a machine to be considered a legitimate moral agent? And what would extending agency to machines mean for our understanding of technology, ourselves, and ethics?

1.2   Agency

To address these questions, we first need to define or at least characterize what is meant by the term "moral agent." To get at this, we can begin with what Kenneth Einar Himma (2009, 19) calls "the standard view" of moral agency. We begin here not because this particular conceptualization is necessarily correct and beyond critical inquiry, but because it provides a kind of baseline for the ensuing investigation and an easily recognizable point of departure. Such a beginning, as Hegel (1987) was well aware, is never absolute or without its constitutive presumptions and prejudices. It is always something of a strategic decision—literally a cut into the matter—that will itself need to be explicated and justified in the course of the ensuing examination. Consequently, we begin not with absolute certainty about what constitutes moral agency but with a standard characterization that will itself need to be investigated and submitted to critical evaluation.

"Moral agency" is, both grammatically and conceptually speaking, a subset of the more general term "agency." Agency, however, is a concept that has a rather specific characterization within the Western philosophical tradition. "The idea of agency," Himma explains, "is conceptually associated with the idea of being capable of doing something that counts as an act or action. As a conceptual matter, X is an agent if and only if X is capable of performing action. Actions are doings, but not every doing is an action; breathing is something we do, but it does not count as an action. Typing these words is an action, and it is in virtue of my ability to do this kind of thing that, as a conceptual matter, I am an *agent*" (Himma 2009, 19–20). Furthermore, agency, at least as it is typically characterized and understood, requires that there be some kind of animating "intention" behind the observed action. "The difference between breathing and typing words," Himma continues, "is that the latter depends on my having a certain kind of mental state" (ibid., 20). In this way, agency can be explained by way of what Daniel Dennett calls an "intentional system," which is characterized as any system—be it a man, machine, or alien creature (Dennett 1998, 9)—to which one can ascribe "beliefs and desires" (Dennett 1998, 3). Consequently, "only beings capable of intentional states (i.e., mental states that are about something else, like a desire for X), then, are agents. People and dogs are both capable of performing acts because both are capable of intentional states. . . . In contrast, trees are not agents, at bottom, because trees are incapable of intentional states (or any other mental state, for that matter). Trees grow leaves, but growing leaves is not something that happens as the result of an action on the part of the tree"

(Himma 2009, 20). For this reason, agency is something that tends to be restricted to human individuals and animals—entities that can have intentions to act and that can, on the basis of that, perform an action. Everything else, like plants, rocks, and other inanimate objects, would be located outside the realm of agency. Although actions might and often do involve these other kinds of entities, they are not considered agents in their own right. A rock, for instance, might be thrown at a dog by a cruel child. But this mere object (the rock) is not, at least in most circumstances, considered to be responsible or accountable for this action.

A critical demonstration of this widely accepted "fact" is provided in Werner Herzog's cinematic adaptation of the story of Kaspar Hauser, *Every Man for Himself and God against All* (1974). Kaspar, a real historic figure, was a feral child who, according to the historical records, was "kept in a dungeon, separate from all communication with the world, from early childhood to about the age of seventeen" (Von Feuerbach and Johann 1833). According to the plot of Herzog's film, Kaspar (portrayed by the German street musician Bruno S.) is forcefully extracted from his seventeen years of solitary confinement (in a gesture that is reminiscent of the release of the prisoner in Plato's "Allegory of the Cave" from book VII of the *Republic*) and eventually allowed to join human society, but only as a kind of constitutive exception. Because of this "outsider position," Kaspar often provides surprising and paradigm-challenging insights that are, from the perspective of the well-educated men around him, incorrect and erroneous.

In one scene, Kaspar's teacher, Professor Daumer (Walter Ladengast), endeavors to explain how apples ripen on the tree and eventually fall to the ground. In reply to this well-intended explanation, Kaspar suggests that the situation is entirely otherwise. The apples, he opines, actually fall to the ground because "they're tired and want to sleep." The thoughtful and exceedingly patient teacher replies to this "mistaken conclusion" with the following correction: "Kaspar, an apple cannot be tired. Apples do not have lives of their own—they follow our will. I'm going to roll one down the path, it will stop where I want it to." Professor Daumer then rolls an apple down the path, but, instead of stopping at the intended spot in the path, it gets diverted and lands in the grass. Drawing an entirely different conclusion from this occurrence, Kaspar points out that "the apple didn't stop, it hid in the grass." Frustrated by Kaspar's continued lack of understanding, the professor concocts another demonstration, this time enlisting the help of a clergyman, Herr Fuhrmann, who has come to the Daumers' home to evaluate Kaspar's spiritual development. "Now, Herr Fuhrmann," the professor explains, "is going to put out his foot, and, when I roll the apple, it will stop where we want it to." The professor once again rolls the apple, and, instead of stopping as predicted, it bounces over Herr Fuhrmann's foot and down the path. To which Kaspar remarks, "Smart apple! It jumped over his foot and ran away."

The comedic effect of this scene is the product of a conflict between two very different understandings of the attribution of agency. Whereas the professor and the priest *know* that inanimate objects, like apples, are just things that obey our will and only do what we impart to them, Kaspar assigns both agency and sentience to the object. According to Kaspar, it is the apple that does not stop, hides in the grass, and demonstrates intelligence by jumping the obstacle and running away. For the two enlightened men of modern science, however, this conclusion is obviously incorrect and erroneous. For them, agency is something that is, at least in this particular situation, only attributed to human individuals. They know that apples are not, to use Dennett's terminology, "intentional systems."

"Moral agency," as a further qualification and subset of the general category "agency," would include only those agents whose actions are directed by or subject to some moral criteria or stipulation. Understood in this fashion, a dog may be an agent, but it would not be a moral agent insofar as its behavior (e.g., barking at strangers, chasing squirrels, biting the postman) would not be something decided on the basis of, for example, the categorical imperative or some utilitarian calculus of possible outcomes. As J. Storrs Hall (2007, 27) describes it, by way of a somewhat curious illustration, "if the dog brings in porn flyers from the mailbox and gives them to your kids, it's just a dog, and it doesn't know any better. If the butler does it, he is a legitimate target of blame." Although the dog and the butler perform the same action, it is only the butler and not the dog who is considered a moral agent and therefore able to be held accountable for the action. "According to the standard view," Himma (2009, 21) writes, "the concept of moral agency is ultimately a normative notion that picks out the class of beings whose behavior is subject to moral requirements. The idea is that, as a conceptual matter, the behavior of a moral agent is governed by moral standards, while the behavior of something that is not a moral agent is not governed by moral standards."

To be considered a moral agent, therefore, something more is needed beyond what is stipulated for agency in general. Himma again provides a schematic definition derived from a review of the standard account provided in the philosophical literature. "The conditions for moral agency can thus be summarized as follows: for all X, X is a moral agent if and only if X is (1) an agent having the capacities for (2) making free choices, (3) deliberating about what one ought to do, and (4) understanding and applying moral rules correctly in paradigm cases. As far as I can tell, these conditions, though somewhat underdeveloped in the sense that the underlying concepts are themselves in need of a fully adequate conceptual analysis, are both necessary and sufficient for moral agency" (Himma 2009, 24). Articulated in this manner, membership in the community of moral agents will be limited to anything that exhibits or is able to demonstrate the achievement of all four criteria. This means, therefore, that a moral agent, according to Himma's conceptualization of the standard account, is anything capable of performing an intentional action, freely selected through some kind of deliberative process, and directed by following or applying some sort of codified rules.

But this is where things get exceedingly complicated, for a number of reasons. First, this particular characterization of moral agency mobilizes and is supported by metaphysical concepts, like "free choice," "deliberation," and "intentionality"—concepts that are themselves open to considerable debate and philosophical disagreement. To make matters worse, these metaphysical difficulties are further complicated by epistemological problems concerning access to and the knowability of another individual's inner dispositions, or what philosophers routinely call "the other minds problem." That is, if an agent's actions were, for example, freely chosen (whatever that might mean) through some kind of deliberation (whatever that might refer to), how would this kind of activity be accessed, assessed, or otherwise available to an outside observer? How, in other words, would an agent's "free will" and "deliberation" show itself as such so that one could recognize that something or someone was in fact a moral agent?

Second, because of these complications, this particular characterization of moral agency is not definitive, universal, or final. What Himma calls "the standard account," although providing what is arguably a useful characterization that is "widely accepted" and "taken for granted" (Himma 2009, 19) in much of the current literature, is only one possible definition.

There are many others. As Paul Shapiro (2006, 358) points out, "there are many ways of defining moral agency, and the choice of a definition is a crucial factor in whether moral agency proves to be limited to humans." Himma's investigation begins by explicitly excluding these other, "non-standard" positions, as they would inevitably be called by comparison. "Although there are a number of papers challenging the standard account," Himma (2009, 19) admits, "I will not consider them here." It is with this brief acknowledgment of exclusion—a mark within the text of what has been deliberately left out of consideration within the text—that Himma's essay demonstrates, in very practical terms, how something that is considered to be "the standard account" comes to be standardized. It all comes down to making an exclusive decision, a decisive cut between what is and what is not included. Some things are to be admitted and incorporated into the standard account; everything else, every other possibility, is immediately marginalized as other. Furthermore, these others, precisely because they are not given any further consideration, are only manifest by way of this exclusion. The other, therefore, is only manifest insofar as its marginalization—its othering, if we can be permitted such a term—is marked within the interior of the text as that which is not to be given any further consideration.

Finally, and directly following from this point, providing a definition of "moral agency," as Shapiro's comment indicates and Himma's gesture demonstrates, is not some disinterested and neutral activity. It is itself a decision that has definite moral consequences insofar as a definition—any definition—already and in advance decides who or what should have moral standing and who or what does not. "Philosophers like Pluhar," Shapiro (2006, 358) writes, "set the standard for moral agency at a relatively high level: the capability to understand and act on moral principles. In order to meet this standard, it seems necessary for a being to possess linguistic capabilities beyond those presently ascribed to other species. However, a lower standard for moral agency can also be selected: the capacity for virtuous behavior. If this lower standard is accepted, there can be little doubt that many other animals are moral agents to some degree." In other words, the very act of characterizing moral agency already and unavoidably makes a decision between who or what is included and who or what is to be excluded—who is a member of the club, and who is marginalized as its constitutive other. These decisions and their resultant exclusions, as even a brief survey of the history of moral philosophy demonstrates, often have devastating consequences for others. At one time, for example, the standard for "moral agency" was defined in such a way that it was limited to white European males. This obviously had significant material, legal, and ethical consequences for all those others who were already excluded from participating in this exclusive community—women, people of color, non-Europeans, and so on. Consequently, what matters in any investigation of this subject matter is not only who is and who is not considered a moral agent, but also, and perhaps more importantly, how one first defines "moral agency," who or what gets to decide these things, on what grounds, and with what outcomes.

1.3   The Mechanisms of Exclusion

Computers and related systems are typically understood to be and conceptualized as a tool. "I believe," John Searle (1997, 190) writes, "that the philosophical importance of computers, as is typical with any new technology, is grossly exaggerated. The computer is a useful tool, nothing more nor less." Questions of agency and especially moral agency are, on this account, situated not in the materiality of the instrument but in its design, use, or implementation. "The moral value of purely mechanical objects," as David F. Channell (1991, 138) explains, "is determined by factors that are external to them—in effect, by the usefulness to human beings." It is, therefore, not the tool but the human designer and/or user who is accountable for any and all actions involving the device. This decision is structured and informed by the answer that is typically provided for the question concerning technology. "We ask the question concerning technology," Martin Heidegger (1977a, 4–5) writes, "when we ask what it is. Everyone knows the two statements that answer our question. One says: Technology is a means to an end. The other says: Technology is a human activity. The two definitions of technology belong together. For to posit ends and procure and utilize the means to them is a human activity. The manufacture and utilization of equipment, tools, and machines, the manufactured and used things themselves, and the needs and ends that they serve, all belong to what technology is." According to Heidegger's analysis, the presumed role and function of any kind of technology, whether it be the product of handicraft or industrialized manufacture, is that it is a means employed by human users for specific ends. Heidegger (ibid., 5) terms this particular characterization of technology "the instrumental definition" and indicates that it forms what is considered to be the "correct" understanding of any kind of technological device.

As Andrew Feenberg (1991, 5) summarizes it in the introduction to his *Critical Theory of Technology*, "the instrumentalist theory offers the most widely accepted view of technology. It is based on the common sense idea that technologies are 'tools' standing ready to serve the purposes of users." And because an instrument "is deemed 'neutral,' without valuative content of its own" (ibid.), a technological artifact is evaluated not in and of itself, but on the basis of the particular employments that have been decided by its human designer or user. This verdict is succinctly summarized by Jean-François Lyotard in *The Postmodern Condition*: "Technical devices originated as prosthetic aids for the human organs or as physiological systems whose function it is to receive data or condition the context. They follow a principle, and it is the principle of optimal performance: maximizing output (the information or modification obtained) and minimizing input (the energy expended in the process). Technology is therefore a game pertaining not to the true, the just, or the beautiful, etc., but to efficiency: a technical 'move' is 'good' when it does better and/or expends less energy than another" (Lyotard 1984, 44). Lyotard's explanation begins by affirming the traditional understanding of technology as an instrument, prosthesis, or extension of human faculties. Given this "fact," which is stated as if it were something that is beyond question, he proceeds to provide an explanation of the proper place of the technological apparatus in epistemology, ethics, and aesthetics. According to his analysis, a technological device, whether it be a cork screw, a clock, or a computer, does not in and of itself participate in the big questions of truth, justice, or beauty. Technology, on this account, is simply and indisputably about efficiency. A particular technological innovation is considered "good," if, and only if, it proves to be a more effective means to accomplishing a desired end.

The instrumentalist definition is not merely a matter of philosophical refection, it also informs and serves as the conceptual backdrop for work in AI and robotics, even if it is often not identified as such.[1] Joanna Bryson, for instance, mobilizes the instrumentalist perspective in her essay "Robots Should Be Slaves." "Legal and moral responsibility for a robot's actions should be no different than they are for any other AI system, and these are the same as for any other tool. Ordinarily, damage caused by a tool is the fault of the operator, and benefit from it is to the operator's credit. . . . We should never be talking about machines taking ethical decisions, but rather machines operated correctly within the limits we set for them" (Bryson 2010, 69). For Bryson, robots and AI systems are no different from any other technical artifact. They are tools of human manufacture, employed by human users for particular purposes, and as such are merely "an extension of the user" (ibid., 72). Bryson, therefore, would be in agreement with Marshall McLuhan, who famously characterized all technology as media—literally the means of effecting or conveying—and all media as an extension of human faculties. This is, of course, immediately evident from the title of what is considered McLuhan's most influential book, *Understanding Media: The Extensions of Man*. And the examples used throughout this text are by now familiar: the wheel is an extension of the foot, the telephone is an extension of the ear, and television is an extension of the eye (McLuhan 1995). Conceptualized in this fashion, technical mechanisms are understood as *prostheses* through which various human facilities come to be extended beyond their original capacity or natural ability.

In advancing this position, McLuhan does not so much introduce a new understanding of technology as he provides explicit articulation of a decision that is itself firmly rooted in the soil of the Western tradition. The concept of technology, especially the technology of information and communication, as an extension of human capabilities is already evident in Plato's *Phaedrus*, where writing had been presented and debated as an artificial supplement for speech and memory (Plato 1982, 274b–276c). And Socrates is quite clear on this point: writing is just a tool, it means nothing by itself. It only says one and the same thing. As if repeating this Socratic evaluation, John Haugeland argues that artifacts "only have meaning because we give it to them; their intentionality, like that of smoke signals and writing, is essentially borrowed, hence *derivative*. To put it bluntly: computers themselves don't mean anything by their tokens (any more than books do)—they only mean what we say they do. Genuine understanding, on the other hand, is intentional 'in its own right' and not derivatively from something else" (Haugeland 1981, 32–33). Dennett explains this position by considering the example of an encyclopedia. Just as the printed encyclopedia is a reference tool for human users, so too would be an automated computerized encyclopedia. Although interacting with such a system, like Wikipedia, might give the impression that one was "communicating with another person, another entity endowed with original intentionality," it is "still just a tool, and whatever meaning or aboutness we vest in it is just a byproduct of our practices in using the device to serve our own goals" (Dennett 1989, 298).

Understood as an extension or enhancement of human faculties, sophisticated technical devices like robots, AIs, and other computer systems are not considered the responsible agent of actions that are performed with or through them. "Morality," as Hall (2001, 2) points out, "rests on human shoulders, and if machines changed the ease with which things were done, they did not change responsibility for doing them. People have always been the only 'moral agents.'" This formulation not only sounds level-headed and reasonable, it is one of the standard operating presumptions of computer ethics. Although different definitions of "computer ethics" have circulated since Walter Maner first introduced the term in 1976 (Maner 1980), they all share a human-centered perspective that assigns moral agency to human designers and users. According to Deborah Johnson, who is credited with writing the field's agenda-setting textbook, "computer ethics turns out to be the study of human beings and society— our goals and values, our norms of behavior, the way we organize ourselves and assign rights and responsibilities, and so on" (Johnson 1985, 6). Computers, she recognizes, often "instrumentalize" these human values and behaviors in innovative and challenging ways, but the bottom line is and remains the way human agents design and use (or misuse) such technology. And Johnson has stuck to this conclusion even in the face of what appears to be increasingly sophisticated technological developments. "Computer systems," she writes in a more recent article, "are produced, distributed, and used by people engaged in social practices and meaningful pursuits. This is as true of current computer systems as it will be of future computer systems. No matter how independently, automatic, and interactive computer systems of the future behave, they will be the products (direct or indirect) of human behavior, human social institutions, and human decision" (Johnson 2006, 197). Understood in this way, computer systems, no matter how automatic, independent, or seemingly intelligent they may become, "are not and can never be (autonomous, independent) moral agents" (ibid., 203). They will, like all other technological artifacts, always be instruments of human value, decision making, and action.[2]

According to the instrumentalist definition, therefore, any action undertaken by a computer system is ultimately the responsibility of some human agent—the designer of the system, the manufacturer of the equipment, or the end-user of the product. If something goes wrong or someone is harmed by the mechanism, "some human is," as Ben Goertzel (2002, 1) accurately describes it, "to blame for setting the program up to do such a thing." Following this line of argument, therefore, the "death by robot" scenarios with which we began would ultimately be the fault of some human programmer, manufacturer, or operator. Holding the robotic mechanism or AI system culpable would be, on this account, not only absurd but also irresponsible. Ascribing moral agency to machines, Mikko Siponen argues, allows one to "start blaming computers for our mistakes. In other words, we can claim that 'I didn't do it—it was a computer error,' while ignoring the fact that the software has been programmed by people to 'behave in certain ways,' and thus people may have caused this error either incidentally or intentionally (or users have otherwise contributed to the cause of this error)" (Siponen 2004, 286). This line of thinking has been codified in the popular adage, "It's a poor carpenter who blames his tools." In other words, when something goes wrong or a mistake is made in situations involving the application of technology, it is the operator of the tool and not the tool itself that should be blamed. Blaming the tool is not only ontologically incorrect, insofar as a tool is just an extension of human action, but also ethically suspect, because it is one of the ways that human agents often try to deflect or avoid taking responsibility for their actions.

For this reason, researchers caution against assigning moral agency to machines not only because doing so is conceptually wrong or disputed, but also because it gives human beings license to blame their tools. "By endowing technology with the attributes of autonomous agency," Abbe Mowshowitz (2008, 271) argues, "human beings are ethically sidelined. Individuals are relieved of responsibility. The suggestion of being in the grip of irresistible forces provides an excuse of rejecting responsibility for oneself and others." This maneuver, what Helen Nissenbaum (1996, 35) terms "the computer as scapegoat," is understandable but problematic, insofar as it complicates moral accountability, whether intentionally or not:

Most of us can recall a time when someone (perhaps ourselves) offered the excuse that it was the computer's fault—the bank clerk explaining an error, the ticket agent excusing lost bookings, the student justifying a late paper. Although the practice of blaming a computer, on the face of it, appears reasonable and even felicitous, it is a barrier to accountability because, having found one explanation for an error or injury, the further role and responsibility of human agents tend to be underestimated—even sometimes ignored. As a result, no one is called upon to answer for an error or injury. (Ibid.)

And it is precisely for this reason that Johnson and Miller (2008, 124) argue that "it is dangerous to conceptualize computer systems as autonomous moral agents."

The instrumental theory not only sounds reasonable, it is obviously useful. It is, one might say, instrumental for parsing the question of agency in the age of increasingly complex technological systems. And it has the advantage that it situates moral responsibility in a widely accepted and seemingly intuitive subject position, in human decision making and action, and resists any and all efforts to defer responsibility to some inanimate object by blaming what are mere instruments or tools. At the same time, however, this particular formulation also has significant theoretical and practical limitations. Theoretically, it is an anthropocentric theory. As Heidegger (1977a) pointed out, the instrumental definition of technology is conceptually tethered to an assumption concerning the position and status of the human being. Anthropocentrism, however, has at least two problems.

First, the concept "human" is not some eternal, universal, and immutable Platonic idea. In fact, who is and who is not "human" is something that has been open to considerable ideological negotiations and social pressures. At different times, membership criteria for inclusion in club *anthropos* have been defined in such a way as to not only exclude but to justify the exclusion of others, for example, barbarians, women, Jews, and people of color. This "sliding scale of humanity," as Joanna Zylinska (2009, 12) calls it, institutes a metaphysical concept of the human that is rather inconsistent, incoherent, and capricious. As membership in the club has slowly and not without considerable resistance been extended to these previously excluded populations, there remain other, apparently more fundamental, exclusions, most notably that of nonhuman animals and technological artifacts. But even these distinctions are contested and uncertain. As Donna Haraway has argued, the boundaries that had once neatly separated the concept of the human from its traditionally excluded others have broken down and become increasingly untenable:

By the late twentieth century in United States, scientific culture, the boundary between human and animal is thoroughly breached. The last beachheads of uniqueness have been polluted, if not turned into amusement parks—language, tool use, social behavior, mental events. Nothing really convincingly settles the separation of human and animal. [Additionally] late twentieth century machines have made thoroughly ambiguous the difference between natural and artificial, mind and body, self-developing and externally designed, and many other distinctions that used to apply to organisms and machines. Our machines are disturbingly lively, and we ourselves frighteningly inert. (Haraway 1991, 151–152)

Second, anthropocentrism, like any centrism, is exclusive. Such efforts draw a line of demarcation and decide who is to be considered an insider and who is to be identified as an outsider. The problem, however, is not only who gets to draw the line and what comprises the criterion of inclusion, the problem is that this operation, irrespective of the specific criteria that come to be applied, is by definition violent and exclusive. "The institution of *any* practice of *any* criterion of moral considerability," Thomas Birch (1993, 317) writes, "is an act of power over, and ultimately an act of violence toward, those others who turn out to fail the test of the criterion and are therefore not permitted to enjoy the membership benefits of the club of *consideranda*. They become 'fit objects' of exploitation, oppression, enslavement, and finally extermination. As a result, the very question of moral considerability is ethically problematic itself, because it lends support to the monstrous Western project of planetary domination." Consequently, the instrumental theory, which proposes that all technology be considered neutral and in itself beyond good and evil, is not a neutral instrument. It already is part of and participates in an "imperial program" (ibid., 316) that not only decides who should be considered a proper moral subject but also, and perhaps worse, legitimates the use and exploitation of others.

The example typically utilized to illustrate this point is animal research and testing. Because animals are determined to be otherwise than human, they are able to be turned into instruments of and for human knowledge production. Although the violence visited upon these others and even their eventual death is regrettable, it is, so the argument goes, a means to a higher (humanly defined) end. For this reason, human beings working in the field of animal rights philosophy argue that the real culprit, the proverbial "bad guy," in these situations is anthropocentrism itself. As Matthew Calarco argues, "the genuine critical target of progressive thought and politics today should be *anthropocentrism* as such, for it is always one version or another of *the human* that falsely occupies the space of the universal and that functions to exclude what is considered nonhuman (which, of course, includes the immense majority of human beings themselves, along with all else deemed to be nonhuman) from ethical and political consideration" (Calarco 2008, 10). The main theoretical problem with the instrumental definition of technology, then, is that it leaves all of this uninterrogated and in doing so not only makes potentially inaccurate ontological decisions about who is and who is not included but also risks enacting moral decisions that have potentially devastating consequences for others.

Practically, the instrumental theory succeeds only by reducing technology, irrespective of design, construction, or operation, to a tool—a prosthesis or extension of human agency. "Tool," however, does not necessarily encompass everything technological and does not exhaust all possibilities. There are also *machines*. Although "experts in mechanics," as Karl Marx (1977, 493) pointed out, often confuse these two concepts, calling "tools simple machines and machines complex tools," there is an important and crucial difference between the two, and that difference ultimately has to do with the location and assignment of agency. An indication of this essential difference can be found in a brief parenthetical aside offered by Heidegger in "The Question Concerning Technology." "Here it would be," Heidegger (1977a, 17) writes in reference to his use of the word "machine" to characterize a jet airliner, "to discuss Hegel's definition of the machine as autonomous tool [*selbständigen Werkzeug*]." What Heidegger references, without supplying the full citation, are Hegel's 1805–1807 Jena Lectures, in which "machine" had been defined as a tool that is self-sufficient, self-reliant, or independent. Although Heidegger immediately dismisses this alternative as something that is not appropriate to his way of questioning technology, it is taken up and given sustained consideration by Langdon Winner in *Autonomous Technology*:

To be autonomous is to be self-governing, independent, not ruled by an external law of force. In the metaphysics of Immanuel Kant, autonomy refers to the fundamental condition of free will—the capacity of the will to follow moral laws which it gives to itself. Kant opposes this idea to "heteronomy," the rule of the will by external laws, namely the deterministic laws of nature. In this light the very mention of autonomous technology raises an unsettling irony, for the expected relationship of subject and object is exactly reversed. We are now reading all of the propositions backwards. To say that technology is autonomous is to say that it is nonheterono-mous, not governed by an external law. And what is the external law that is appropriate to technology? Human will, it would seem. (Winner 1977, 16)

"Autonomous technology," therefore, refers to technical devices that directly contravene the instrumental definition by deliberately contesting and relocating the assignment of agency. Such mechanisms are not mere tools to be used by human agents but occupy, in one way or another, the place of agency. As Marx (1977, 495) described it, "the machine, therefore, is a mechanism that, after being set in motion, performs with its tools the same operations as the worker formerly did with similar tools." Understood in this way, the machine replaces not the hand tool of the worker but the worker him- or herself, the active agent who had wielded the tool.

The advent of autonomous technology, therefore, introduces an important conceptual shift that will have a significant effect on the assignment and understanding of moral agency. "The 'artificial intelligence' programs in practical use today," Goertzel (2002, 1) admits, "are sufficiently primitive that their morality (or otherwise) is not a serious issue. They are intelligent, in a sense, in narrow domains—but they lack autonomy; they are operated by humans, and their actions are integrated into the sphere of human or physical-world activities directly via human actions. If such an AI program is used to do something immoral, some human is to blame for setting the program up to do such a thing." In stating this, Goertzel, it seems, would be in complete agreement with instrumentalists like Bryson, Johnson, and Nissenbaum insofar as current AI technology is still, for the most part, under human control and therefore able to be adequately explained and conceptualized as a mere tool. But that will not, Goertzel argues, remain for long. "Not too far in the future, however, things are going to be different. AI's will possess true artificial general intelligence (AGI), not necessarily emulating human intelligence, but equaling and likely surpassing it. At this point, the morality or otherwise of AGI's will become a highly significant issue" (ibid.).

This kind of forecasting is shared and supported by other adherents of the "hard take-off hypothesis" or "singularity thesis." Celebrated AI scientists and robotics researchers like Ray Kurzweil and Hans Moravec have issued similar optimistic predictions. According to Kurzweil's (2005, 8) estimations, technological development is "expanding at an exponential pace," and, for this reason, he proposes the following outcome: "within several decades information-based technologies will encompass all human knowledge and proficiency, ultimately including the pattern recognition powers, problem solving skills, and emotional and moral intelligence of the human brain itself" (ibid.). Similarly, Hans Moravec forecasts not only the achievement of human-level intelligence in a relatively short period of time but an eventual surpassing of it that will render human beings effectively obsolete and a casualty of our own evolutionary progress.

We are very near to the time when virtually no essential human function, physical or mental, will lack an artificial counterpart. The embodiment of this convergence of cultural developments will be the intelligent robot, a machine that can think and act as a human, however inhuman it may be in physical or mental detail. Such machines could carry on our cultural evolution, including their own construction and increasingly rapid self-improvement, without us, and without the genes that built us. When that happens, our DNA will find itself out of a job, having lost the evolutionary race to a new kind of competition. (Moravec 1988, 2)

Even seemingly grounded and level-headed engineers, like Rodney Brooks, who famously challenged Moravec and the AI establishment with his "mindless" but embodied and situated robots, predict the achievement of machine intelligence on par with humanlike capabilities in just a few decades. "Our fantasy machines," Brooks (2002, 5) writes, referencing the popular robots of science fiction (e.g., HAL, C-3PO, Lt. Commander Data), "have syntax and technology. They also have emotions, desires, fears, loves, and pride. Our real machines do not. Or so it seems at the dawn of the third millennium. But how will it look a hundred years from now? My thesis is that in just twenty years the boundary between fantasy and reality will be rent asunder."

If these predictions are even partially correct and accurate, then what has been defined and largely limited to the status of a mere instrument will, at some point in the not too distant future, no longer be just a tool or an extension of human capabilities. What had been considered a tool will be as intelligent as its user, if not capable of exceeding the limits of human intelligence altogether. If this prediction turns out to have traction and we successfully fashion, as Kurzweil (2005, 377) predicts, "nonbiological systems that match and exceed the complexity and subtlety of humans, including our emotional intelligence," then continuing to treat such artifacts as mere instruments of our will would be not only be terribly inaccurate but also, and perhaps worse, potentially immoral. "For all rational beings," irrespective of origin or composition, Kant argues in the *Grounding for the Metaphysics of Morals* (1983, 39), "stand under the law that each of them should treat himself and all others never merely as a means but always at the same time as an end in himself."[3] Following this line of argument, we can surmise that if AIs or robots were capable of achieving an appropriate level of rational thought, then such mechanisms will be and should be included in what Kant (1983, 39) termed "the kingdom of ends." In fact, barring such entities from full participation in this "systemic union of rational beings" (ibid.) and continuing to treat the *machina ratiocinatrix*, as Norbert Wiener (1996, 12) called it, as a mere means to be controlled and manipulated by another, is typically the motivating factor for the "robots run amok" or "machine rebellion" scenario that is often portrayed in science fiction literature and film. .

These narratives typically play out in one of two ways. On the one hand, human beings become, as Henry David Thoreau (1910, 41) once described it, "the tool of our tools." This dialectical inversion of user and tool or master and slave, as it was so insightfully demonstrated in Hegel's 1807 *Phenomenology of Spirit*, is dramatically illustrated in what is perhaps the most popular science fiction franchise from the turn of the twenty-first century—*The Matrix*. According to the first episode of the cinematic trilogy (*The Matrix*, 1999), the computers win a struggle for control over their human masters and turn the surviving human population into a bio-electrical power supply source to feed the machines. On the other hand, our technological creations, in a perverse version of Moravec's (1988) prediction, rise up and decide to dispense with humanity altogether. This scenario often takes the dramatic form of violent revolution and even genocide. In Čapek's *R.U.R.*, for instance, the robots, in what many critics consider to be a deliberate reference to the workers' revolutions of the early twentieth century, begin seeding revolt by printing their own manifesto: "Robots of the world! We the first union at Rossum's Universal Robots, declare that man is our enemy and the blight of the universe . . . Robots of the world, we enjoin you to exterminate mankind. Don't spare the men. Don't spare the women. Retain all factories, railway lines, machines and equipment, mines and raw materials. All else should be destroyed" (Čapek 2008, 67). A similar apocalyptic tone is deployed in Ron Moore's reimagined version of *Battlestar Galactica* (2003–2009). "The cylons were created by man," the program's tag-line refrain read. "They rebelled. They evolved. And they have a plan." That plan, at least as it is articulated in the course of the miniseries, appears to be nothing less than the complete annihilation of the human race.

Predictions of human-level (or better) machine intelligence, although fueling imaginative and entertaining forms of fiction, remain, for the most part, futuristic. That is, they address possible achievements in AI and robotics that might occur with technologies or techniques that have yet to be developed, prototyped, or empirically demonstrated. Consequently, strict instrumentalists, like Bryson or Johnson, are often able to dismiss these prognostications of autonomous technology as nothing more than wishful thinking or speculation. And if the history of AI is any indication, there is every reason to be skeptical. We have, in fact, heard these kinds of fantastic hypotheses before, only to be disappointed time and again. As Terry Winograd (1990, 167) wrote in an honest assessment of progress (or lack thereof) in the discipline, "indeed, artificial intelligence has not achieved creativity, insight, and judgment. But its shortcomings are far more mundane: we have not yet been able to construct a machine with even a modicum of common sense or one that can converse on everyday topics in ordinary language."

Despite these shortcomings, however, there are current implementations and working prototypes that appear to be independent and that complicate the assignment of agency. There are, for instance, autonomous learning systems, mechanisms not only designed to make decisions and take real-world actions with little or no human direction or oversight but also programmed to be able to modify their own rules of behavior based on results from such operations. Such machines, which are now rather common in commodities trading, transportation, health care, and manufacturing, appear to be more than mere tools. Although the extent to which one might assign "moral agency" to these mechanisms is a contested issue, what is not debated is the fact that the rules of the game have changed significantly. As Andreas Matthias points out, summarizing his survey of learning automata:

Presently there are machines in development or already in use which are able to decide on a course of action and to act without human intervention. The rules by which they act are not fixed during the production process, but can be changed during the operation of the machine, *by the machine itself*. This is what we call machine learning. Traditionally we hold either the operator/manufacture of the machine responsible for the consequences of its operation or "nobody" (in cases where no personal fault can be identified). Now it can be shown that there is an increasing class of machine actions, where the traditional ways of responsibility ascription are not compatible with our sense of justice and the moral framework of society because nobody has enough *control* over the machine's actions to be able to assume responsibility for them. (Matthias 2004, 177)

In other words, the instrumental definition of technology, which had effectively tethered machine action to human agency, no longer applies to mechanisms that have been deliberately designed to operate and exhibit some form, no matter how rudimentary, of independent or autonomous action.[4] This does not mean, it is important to emphasize, that the instrumental definition is on this account refuted *tout court*. There are and will continue to be mechanisms understood and utilized as tools to be manipulated by human users (e.g., lawnmowers, cork screws, telephones, digital cameras). The point is that the instrumentalist definition, no matter how useful and seemingly correct it may be in some circumstances for explaining some technological devices, does not exhaust all possibilities for all kinds of technology.

In addition to sophisticated learning automata, there are also everyday, even mundane examples that, if not proving otherwise, at least significantly complicate the instrumentalist position. Miranda Mowbray, for instance, has investigated the complications of moral agency in online communities and massive multiplayer online role playing games (MMORPGs or MMOs):

The rise of online communities has led to a phenomenon of real-time, multiperson interaction via online personas. Some online community technologies allow the creation of bots (personas that act according to a software programme rather than being directly controlled by a human user) in such a way that it is not always easy to tell a bot from a human within an online social space. It is also possible for a persona to be partly controlled by a software programme and partly directed by a human. . . . This leads to theoretical and practical problems for ethical arguments (not to mention policing) in these spaces, since the usual one-to-one correspondence between actors and moral agents can be lost. (Mowbray 2002, 2)

Software bots, therefore, not only complicate the one-to-one correspondence between actor and moral agent but make it increasingly difficult to decide who or what is responsible for actions in the virtual space of an online community.

Although bots are by no means the kind of AGI that Goertzel and company predict, they can still be mistaken for and pass as other human users. This is not, Mowbray points out, "a feature of the sophistication of bot design, but of the low bandwidth communication of the online social space" (ibid.), where it is "much easier to convincingly simulate a human agent." To complicate matters, these software agents, although nowhere near to achieving anything that looks remotely like human-level intelligence, cannot be written off as mere instruments or tools. "The examples in this paper," Mowbray concludes, "show that a bot may cause harm to other users or to the community as a whole by the will of its programmers or other users, but that it also may cause harm through nobody's fault because of the combination of circumstances involving some combination of its programming, the actions and mental/emotional states of human users who interact with it, behavior of other bots and of the environment, and the social economy of the community" (ibid., 4). Unlike AGI, which would occupy a position that would, at least, be on par with that of a human agent and therefore not be able to be dismissed as a mere tool, bots simply muddy the waters (which is probably worse) by leaving undecided the question of whether they are or are not tools. And in the process, they leave the question of moral agency both unsettled and unsettling.

From a perspective that already assumes and validates the instrumental definition, this kind of artificial autonomy, whether manifest in the form of human-level or better AGI or the seemingly mindless operations of software bots, can only be registered and understood as a loss of control by human agents over their technological artifacts. For this reason, Winner initially defines "autonomous technology" negatively. "In the present discussion, the term *autonomous technology* is understood to be a general label for all conceptions and observations to the effect that technology is somehow out of control by human agency" (Winner 1977, 15). This "technology out of control" formulation not only has considerable pull in science fiction, but also fuels a good deal of work in modern literature, social criticism, and political theory. And Winner marshals an impressive roster of thinkers and writers who, in one way or another, worry about and/or criticize the fact that our technological devices not only exceed our control but appear to be in control of themselves, if not threatening to take control of us. Structured in this clearly dramatic and antagonistic fashion, there are obvious winners and losers. In fact, for Jacques Ellul, who is Winner's primary source for this material, "technical autonomy" and "human autonomy" are fundamentally incompatible and mutually exclusive (Ellul 1964, 138). For this reason, Winner ends his *Autonomous Technology* in the usual fashion, with an ominous warning and ultimatum: "Modern people have filled the world with the most remarkable array of contrivances and innovations. If it now happens that these works cannot be fundamentally reconsidered and reconstructed, humankind faces a woefully permanent bondage to the power of its own inventions. But if it is still thinkable to dismantle, to learn and start again, there is a prospect of liberation" (Winner 1977, 335). The basic contours of this story are well known and have been rehearsed many times: some technological innovation has gotten out of control, it now threatens us and the future of humanity, and we need to get it back under our direction, if we are to survive.

This plot line, despite its popularity, is neither necessary nor beyond critical inquiry. In fact, Winner, early in his own analysis, points to another possibility, one that he does not pursue but which nevertheless provides an alternative transaction and outcome: "The conclusion that something is 'out of control' is interesting to us only insofar as we expect that it ought to be in control in the first place. Not all cultures, for example, share our insistence that the ability to control things is a necessary prerequisite of human survival" (Winner 1977, 19). In other words, technology can only be "out of control" and in need of a substantive reorientation or reboot if we assume that it should be under our control in the first place. This assumption, which obviously is informed and supported by an unquestioned adherence to the instrumental definition, already makes crucial and perhaps prejudicial decisions about the ontological status of the technological object. Consequently, instead of trying to regain control over a supposed "tool" that we assume has gotten out of our control or run amok, we might do better to question the very assumption with which this line of argument begins, namely, that these technological artifacts are and should be under our control. Perhaps things can be and even should be otherwise. The critical question, therefore, might not be "how can we reestablish human dignity and regain control of our machines?" Instead we might ask whether there are other ways to address this apparent "problem"—ways that facilitate critical evaluation of the presumptions and legacy of this human exceptionalism, that affirm and can recognize alternative configurations of agency, and that are open to and able to accommodate others, and other forms of otherness.

1.4   The Mechanisms of Inclusion

One way of accommodating others is to define moral agency so that it is neither speciesist nor specious. As Peter Singer (1999, 87) points out, "the biological facts upon which the boundary of our species is drawn do not have moral significance," and to decide questions of moral agency on this ground "would put us in the same position as racists who give preference to those who are members of their race." Toward this end, the question of moral agency has often been referred to and made dependent upon the concept of "personhood." "There appears," G. E. Scott (1990, 7) writes, "to be more unanimity as regards the claim that in order for an individual to be a moral agent s/he must possess the relevant features of a person; or, in other words, that being a person is a necessary, if not sufficient, condition for being a moral agent." In fact, it is on the basis of personhood that other entities have been routinely excluded from moral consideration. As David McFarland asserts:

To be morally responsible, an agent—that is the person performing or failing to perform the function in question—has as a rule a moral obligation, and so is worthy of either praise or blame. The person can be the recipient of what are sometimes called by philosophers their "desert." But a robot is not a person, and for a robot to be given what it deserves—"its just deserts"—it would have to be given something that mattered to it, and it would have to have some understanding of the significance of this. In short, it would have to have some sense of its own identity, some way of realising that *it* was deserving of something, whether pleasant or unpleasant. (McFarland 2008, ix)

The concept *person*, although routinely employed to justify and defend decisions concerning inclusion or exclusion, has a complicated history, one that, as Hans Urs von Balthasar (1986, 18) argues, has been given rather extensive treatment in the philosophical literature. "The word 'person,'" David J. Calverley (2008, 525) points out in a brief gloss of this material, "is derived from the Latin word 'persona' which originally referred to a mask worn by a human who was conveying a particular role in a play. In time it took on the sense of describing a guise one took on to express certain characteristics. Only later did the term become coextensive with the actual human who was taking on the persona, and thus become interchangeable with the term 'human.'" This evolution in terminology is something that, according to Marcel Mauss's anthropological investigation in "The Category of the Person" (in Carrithers, Collins, and Lukes 1985), is specifically Western insofar as it is shaped by the institutions of Roman law, Christian theology, and modern European philosophy. The mapping of the concept *person* onto the figure *human*, however, is neither conclusive, universal, nor consistently applied. On the one hand, "person" has been historically withheld from various groups of human beings as a means of subordinating others. "In Roman law," Samir Chopra and Laurence White (2004, 635) point out, "the paterfamilias or free head of the family was the subject of legal rights and obligations on behalf of his household; his wife and children were only indirect subjects of legal rights and his slaves were not legal persons at all." The U.S. Constitution still includes an anachronistic clause defining slaves, or more specifically "those bound to Service for a Term of Years," as three-fifths of a person for the purpose of calculating federal taxes and the appropriation of Congressional seats. And it is current legal practice in U.S. and European law to withhold some aspects of personhood from the insane and mentally deficient.

On the other hand, philosophers, medical ethicists, animal rights activists, and others have often sought to differentiate what constitutes a person from the human being in an effort to extend moral consideration to previously excluded others. "Many philosophers," Adam Kadlac (2009, 422) argues, "have contended that there is an important difference between the concept of a person and the concept of a human being." One such philosopher is Peter Singer. "Person," Singer writes in the book *Practical Ethics* (1999, 87), "is often used as if it meant the same as 'human being.' Yet the terms are not equivalent; there could be a person who is not a member of our species. There could also be members of our species who are not persons." Corporations, for example, are artificial entities that are obviously otherwise than human, yet they are considered legal persons, having rights and responsibilities that are recognized and protected by both national and international law (French 1979).

Likewise, "some philosophers," as Heikki Ikäheimo and Arto Laitinen (2007, 9) point out, "have argued that in the imaginary situation where you and I were to meet previously unknown, intelligent-looking creatures—say, in another solar system—the most fundamental question in our minds would not be whether they are human (obviously, they are not), but, rather, whether they are persons." This is not only a perennial staple of science fiction from *War of the Worlds* and the *Star Trek* franchise to *District 9*; there is an entire area of interstellar law that seeks to define the rights of and responsibilities for alien life forms (Haley 1963). More down-to-earth, animal rights philosophers, and Singer in particular, argue that certain nonhuman animals, like great apes, but also other higher-order mammals, should be considered persons with a legitimate right to continued existence even though they are an entirely different species. Conversely, some members of the human species are arguably less than full persons in both legal and ethical matters. There is, for instance, considerable debate in health care and bioethics as to whether a human fetus or a brain-dead individual in a persistent vegetative state is a person with an inherent "right to life" or not. Consequently, differentiating the category person from that of human not only has facilitated and justified various forms of oppression and exclusion but also, and perhaps ironically, has made it possible to consider others, like nonhuman animals and artificial entities, as legitimate moral subjects with appropriate rights and responsibilities.

It is, then, under the general concept *person* that the community of moral agents can be opened up to the possible consideration and inclusion of nonhuman others. In these cases, the deciding factor for membership in what Birch (1993, 317) calls "the club of *consideranda*" can no longer be a matter of kin identification or genetic makeup, but will be situated elsewhere and defined otherwise. Deciding these things, however, is open to considerable debate as is evident in Justin Leiber's *Can Animals and Machines Be Persons?* This fictional "dialogue about the notion of a person" (Leiber 1985, ix) consists in the imagined "transcript of a hearing before the United Nations Space Administration Commission" and concerns the "rights of persons" for two inhabitants of a fictional space station—a young female chimpanzee named Washoe-Delta (a name explicitly derived from and making reference to the first chimpanzee to learn and use American sign language) and an AI computer called AL (clearly and quite consciously modeled in both name and function on the HAL 9000 computer of *2001: A Space Odyssey*).

The dialogue begins *in medias res*. The space station is beginning to fail and will need to be shut down. Unfortunately, doing so means terminating the "life" of both its animal and machine occupants. In response to this proposal, a number of individuals have protested the decision, asserting that the station not be shut down "because (1) Washoe-Delta and AL 'think and feel' and as such (2) 'are persons,' and hence (3) 'their termination would violate their "rights as persons"'" (Leiber 1985, 4). Leiber's fictional dialogue, therefore, takes the form of a moderated debate between two parties: a complainant, who argues that the chimpanzee and computer are persons with appropriate rights and responsibilities, and a respondent, who asserts the opposite, namely, that neither entity is a person because only "a human being is a person and a person is a human being" (ibid., 6). By taking this particular literary form, Leiber's dialogue demonstrates, following John Locke (1996, 148), that *person* is not just an abstract metaphysical concept but "a forensic term"—one that is asserted, decided, and conferred through legal means.

Despite the fact that Leiber's dialogue is fictional, his treatment of this subject matter has turned out to be rather prescient. In 2007, an animal rights group in Austria, the Association against Animal Factories or Der Verein gegen Tierfabriken (VGT), sought to protect a chimpanzee by securing legal guardianship for the animal in an Austrian court. The chimpanzee, Matthew Hiasl Pan, was captured in Sierra Leone in 1982 and was to be shipped to a research laboratory, but, owing to problems with documentation, eventually ended up in an animal shelter in Vienna. In 2006, the shelter ran into financial difficulties and was in the process of liquidating its assets, which included selling off its stock of animals. "At the end of 2006," as Martin Balluch and Eberhart Theuer (2007, 1) explain, "a person gave a donation of a large sum of money to the president of the animal rights association VGT on the condition that he may only take possession of it if Matthew has been appointed a legal guardian, who can receive this money at the same time, and who can decide what the two together would want to spend the money on. With this contract, VGT's president could argue to have legal standing to start court proceedings for a legal guardian for Matthew. This application was made on 6th February 2007 at the district court in Mödling, Lower Austria."

In the course of making the petition, which was supported by expert testimony from four professors in the fields of law, philosophy, anthropology, and biology, "an argument was put forward that a chimpanzee, and in particular Matthew, is to be considered a person according to Austrian law" (Balluch and Theuer 2007, 1). In making this argument, the petitioners referenced and utilized recent innovations in animal rights philosophy, especially the groundbreaking work of Peter Singer and other "personists" who have successfully advanced the idea that some animals are and should be considered persons (DeGrazia 2006, 49). Crucial to this line of argumentation is a characterization of "person" that does not simply identify it with or make it dependent upon the species *Homo sapiens*. Unfortunately, Austrian civil law code does not provide an explicit definition of "person," and the extant judicial literature, as Balluch and Theuer point out, provides no guidance for resolving the issue. To make matters worse, the court's ruling did not offer a decision on the matter but left the question open and unresolved. The judge initially dismissed the petition on the grounds that the chimpanzee was neither mentally handicapped nor in imminent danger, conditions that are under Austrian law legally necessary in any guardianship petition. The decision was appealed. The appellate judge, however, turned it down on the grounds that the applicants had no legal standing to make an application. As a result, Balluch and Theuer (2007, 1) explain, "she left the question open whether Matthew is a person or not."

Although no legal petition has been made asking a court or legislature to recognize a machine as a legitimate person, there is considerable discussion and debate about this possibility. Beyond Leiber's dialogue, there are a good number of imagined situations in robot science fiction. In Isaac Asimov's *Bicentennial Man* (1976), for instance, the NDR series robot "Andrew" makes a petition to the World Legislature in order to be recognized as and legally declared a person with full human rights. A similar scene is presented in Barrington J. Bayley's *The Soul of the Robot* (1974, 23), which follows the "personal" trials of a robot named Jasperodus:

Jasperodus' voice became hollow and moody. "Ever since my activation everyone I meet looks upon me as a thing, not as a person. Your legal proceedings are based upon a mistaken premise, namely that I am an object. On the contrary, I am a sentient being."

The lawyer looked at him blankly. "I beg your pardon?"

"I am an authentic person; independent and aware."

The other essayed a fey laugh. "Very droll! To be sure, one sometimes encounters robots so clever that one could swear they had real consciousness! However, as is well known . . ."

Jasperodus interrupted him stubbornly. "I wish to fight my case in person. It is permitted for a construct to speak on his own behalf?"

The lawyer nodded bemusedly. "Certainly. A construct may lay before the court any facts having a bearing on his case—or, I should say on *its* case. I will make a note of it." he scribbled briefly. "But if I were you I wouldn't try to tell the magistrate what you just said to me."

And in the "Measure of a Man" episode of *Star Trek: The Next Generation* (1989), the fate of the android Lieutenant Commander Data is adjudicated by a hearing of the Judge Advocate General, who is charged with deciding whether the android is in fact a mere thing and the property of Star Fleet Command or a sentient being with the legal rights of a person. Although the episode ends satisfactorily for Lt. Commander Data and his colleagues, it also leaves the underlying question unanswered: "This case," the judge explains, speaking from the bench, "has dealt with metaphysics, with questions best left to saints and philosophers. I am neither competent, nor qualified, to answer those. I've got to make a ruling—to try to speak to the future. Is Data a machine? Yes. Is he the property of Starfleet? No. We've all been dancing around the basic issue: does Data have a soul? I don't know that he has. I don't know that I have! But I have got to give him the freedom to explore that question himself. It is the ruling of this court that Lieutenant Commander Data has the freedom to choose."

This matter, however, is not something that is limited to fictional court rooms and hearings. It is, as David J. Calverley indicates, a very real and important legal concern: "As non-biological machines come to be designed in ways which exhibit characteristics comparable to human mental states, the manner in which the law treats these entities will become increasingly important both to designers and to society at large. The direct question will become whether, given certain attributes, a non-biological machine could ever be viewed as a 'legal person'" (Calverley 2008, 523). The question Calverley asks does not necessarily proceed from speculation about the future or mere philosophical curiosity. In fact, it is associated with and follows from an established legal precedent. "There is," Peter Asaro (2007, 4) points out, "in the law a relevant case of legal responsibility resting in a non-human, namely corporations. The limited liability corporation is a non-human entity that has been effectively granted legal rights of a person." In the United States this recognition is explicitly stipulated by federal law: "In determining the meaning of any Act of Congress, unless the context indicates otherwise—the words 'person' and 'whoever' include corporations, companies, associations, firms, partnerships, societies, and joint stock companies, as well as individuals" (1 USC sec. 1). According to U.S. law, therefore, "person" is legally defined as applying not only to human individuals but also to nonhuman, artificial entities. In making this stipulation, however, U.S. law, like the Austrian legal system, which had been involved in the case of Matthew Hiasl Pan, does not provide a definition of "person" but merely stipulates which entities are to be considered legal persons. In other words, the letter of the law stipulates who is to be considered a person without defining what constitutes the concept *person*. Consequently, whether the stipulation could in fact be extended to autonomous machines, AIs, or robots remains an intriguing but ultimately unresolved question.

1.4.1 Personal Properties

If anything is certain from the fictional and nonfictional considerations of the concept, it is that the term "person" is important and influential but not rigorously defined and delimited. The word obviously carries a good deal of metaphysical and moral weight, but what it consists in remains unspecified and debatable. "One might well hope," Dennett (1998, 267) writes, "that such an important concept, applied and denied so confidently, would have clearly formulatable necessary and sufficient conditions for ascription, but if it does, we have not yet discovered them. In the end there may be none to discover. In the end we may come to realize that the concept person is incoherent and obsolete." Responses to this typically take the following form: "While one would be hard pressed," Kadlac (2009, 422) writes, "to convince others that monkeys were human beings, on this way of thinking it would be possible to convince others that monkeys were persons. One would simply have to establish conclusively that they possessed the relevant person-making properties." Such a demonstration, as Kadlac anticipates, has at least two dimensions. First, we would need to identify and articulate the "person-making properties" or what Scott (1990, 74) terms the "person schema." We would need, in other words, to articulate what properties make someone or something a person and do so in such a way that is neither capricious nor skewed by anthropocentric prejudice. Second, once standard qualifying criteria for "person" are established, we would need some way to demonstrate or prove that some entity, human or otherwise, possessed these particular properties. We would need some way of testing for and demonstrating the presence of the personal properties in the entity under consideration. Deciding these two things, despite what Kadlac suggests, is anything but "simple."

To begin with, defining "person" is difficult at best. In fact, answers to the seemingly simple and direct question "what is a person?" turn out to be diverse, tentative, and indeterminate. "According to the Oxford Dictionary," Singer (1999, 87) writes, "one of the current meanings of the term is 'a self conscious or rational being.' This sense has impeccable philosophical precedents. John Locke defines a person as 'A thinking intelligent being that has reason and reflection and can consider itself as itself, the same thinking thing, in different times and places.'" Kadlac (2009, 422) follows suit, arguing that in most cases "properties such as rationality and self-consciousness are singled out as person making." For both Singer and Kadlac, then, the defining characteristics of personhood are self-consciousness and rationality. These criteria, as Singer asserts, appear to have an impeccable philosophical pedigree. They are, for instance, not only grounded in historical precedent, for example, Boethius's (1860, 1343c–d) "persona est rationalis naturae individua substantia," but appear to be widely accepted and acknowledged in contemporary usage. Ikäheimo and Laitinen, who come at the question from another direction, make a similar decision: "Moral statuses obviously rest on ontological features in at least two senses. First, it is more or less unanimously accepted by philosophers, and supported by common sense, that our being rational creatures gives us, or makes us deserving of, a special moral status or statuses with regard to each other. Secondly, it is clearly only rational creatures that are capable of claiming for and acknowledging or respecting, moral statuses" (Ikäheimo and Laitinen 2007, 10). What is interesting about this characterization is not only how the term "person" is operationalized, on grounds that are similar to but not exactly the same as those offered by Singer and Kadlac, but also the way the statement hedges its bets—the "more or less" part of the "unanimously accepted," which allows for some significant slippage or wiggle room with regard to the concept.

Other theorists have offered different, although not entirely incompatible, articulations of qualifying criteria. Charles Taylor (1985, 257), for instance, argues that "generally philosophers consider that to be a person in the full sense you have to be an agent with a sense of yourself as agent, a being which can thus make plans for your life, one who also holds values in virtue of which different plans seem better or worse, and who is capable of choosing between them." Christian Smith (2010, 54), who proposes that personhood should be understood as an "emergent property," lists thirty specific capacities ranging from "conscious awareness" through "language use" and "identity formation" to "interpersonal communication and love." And Dennett (1998, 268), in an attempt to sort out these difficulties, suggests that efforts to identify "the necessary and sufficient conditions" for personhood are complicated by the fact that "there seem to be two notions intertwined here." Although formally distinguishing between the metaphysical notion of the person—"roughly, the notion of an intelligent, conscious, feeling agent"—and the moral notion—"roughly, the notion of an agent who is accountable, who has both rights and responsibilities" (ibid.)—Dennett concludes that "there seems to be every reason to believe that metaphysical personhood is a necessary condition of moral personhood" (ibid., 269). What all these characterizations share, despite their variations and differences, is an assumption, presupposition, or belief that the deciding factor is something that is to be found in or possessed by an individual entity. In other words, it is assumed that what makes someone or something a person is some finite number of identifiable and quantifiable "personal properties," understood in both senses of the phrase as something owned by a person and some essential trait or characteristic that comprises or defines what is called a "person." As Charles Taylor (1985, 257) succinctly explains it, "on our normal unreflecting view, all these powers are those of an individual."

To complicate matters, these criteria are themselves often less than rigorously defined and characterized. Take consciousness, for example, which is not just one element among others but a privileged term insofar it appears, in one form or another, on most if not all of the competing lists. This is because consciousness is considered one of the decisive characteristics, dividing between a merely accidental occurrence and a purposeful act that is directed and understood by the individual agent who decides to do it. "Without consciousness," Locke (1996, 146) concludes, "there is no person." Or as Himma (2009, 19) articulates it, with reference to the standard account, "moral agency presupposes consciousness, i.e. the capacity for inner subjective experience like that of pain or, as Nagel puts it, the possession of an internal something-of-which-it-is-to-be and that the very concept of agency presupposes that agents are conscious." Consciousness, in fact, has been one of the principal mechanisms by which human persons have been historically differentiated from both the animal and machinic other. In the *Meditations on First Philosophy*, for example, Descartes (1988, 82) famously discovers and defines himself as "a thing that thinks," or *res cogitans*. This is immediately distinguished from a *res extensa*, an extended being, which not only describes the human body but also the fundamental ontological condition of both animals and machines. In fact, on the Cartesian account, animals are characterized in exclusively mechanical terms, as mere thoughtless automata that act not by intelligence but merely in accordance with the preprogrammed disposition of their constitutive components. "Despite appearances to the contrary," Tom Regan (1983, 3) writes in his critical assessment of the Cartesian legacy, "they are not aware of anything, neither sights nor sounds, smells nor tastes, heat nor cold; they experience neither hunger nor thirst, fear nor rage, pleasure nor pain. Animals are, he [Descartes] observes at one point, like clocks: they are able to do some things better than we can, just as a clock can keep better time; but, like a clock, animals are not conscious."[5]

Likewise, machines, and not just the mechanical clocks of Descartes's era, are situated in a similar fashion. The robots of Čapek's *R.U.R.* are characterized as "having no will of their own. No passions. No hopes. No soul" (Čapek 2008, 28). Or, as Anne Foerst explains it, "when you look at critiques against AI and against the creation of humanoid machines, one thing which always comes up is 'they lack soul.' That's the more religious terminology. The more secular terminology is 'they lack consciousness'" (Benford and Malartre 2007, 162).[6] This concept is given a more scientific expression in AI literature in the form of what Alan Turing initially called "Lady Lovelace's Objection," which is a variant of the instrumentalist argument. "Our most detailed information of Babbage's Analytical Engine," Turing (1999, 50) writes, "comes from a memoir by Lady Lovelace (1842). In it she states, 'The Analytical Engine has no pretensions to *originate* anything. It can do *whatever we know how to order it* to perform' (her italics)." This objection is often deployed as the basis for denying consciousness to computers, robots, and other autonomous machines. Such machines, it is argued, only do what we have programmed them to do. They are, strictly speaking, thoughtless instruments that make no original decisions or determinations of their own. "As impressive as the antics of these artefacts are,"

Pentti Haikonen (2007, 1) argues, "their shortcoming is easy to see: the lights may be on, but there is 'nobody' at home. The program-controlled microprocessor and the robots themselves do not know what they are doing. These robots are no more aware of their own existence than a cuckoo clock on a good day." For this reason, thoughtful people like Dennett (1994, 133) conclude that "it is unlikely, in my opinion, that anyone will ever make a robot that is conscious in just the way we human beings are."

These opinions and arguments, however, are contested and for a number of reasons. It has, on the one hand, been argued that animals are not simply unconscious, stimulus-response mechanisms, like thermostats or clocks, but have some legitimate claim to mental states and conscious activity. Tom Regan, for example, builds a case for animal rights by directly disputing the Cartesian legacy and imputing consciousness to animals. "There is," Regan (1983, 28) writes, "no one *single* reason for attributing consciousness or a mental life to certain animals. What we have is a *set* of reasons, which when taken together, provides what might be called the *Cumulative Argument for animal consciousness*." The "cumulative argument," as Regan characterizes it, consists in the following five elements: "the commonsense view of the world"; linguistic habits by which conscious mind states often come to be attributed to animals (e.g., Fido is hungry); the critique of human exceptionalism and anthropocentrism that disputes the "strict dichotomy between humans and animals"; animal behavior, which appears to be consciously directed and not generated randomly; and evolutionary theory, which suggests that the difference between animals and humans beings "is one of degree and not of kind" (ibid., 25–28). According to Regan, therefore, "those who refuse to recognize the reasonableness of viewing many other animals, in addition to *Homo sapiens*, as having a mental life are the ones who are prejudiced, victims of human chauvinism—the conceit that we (humans) are *so* very special that we are the only conscious inhabitants on the face of the earth" (ibid., 33). The main problem for Regan, however, is deciding which animals qualify as conscious entities and which do not. Although Regan recognizes that "*where one draws the line* regarding the presence of consciousness is not an easy matter" (ibid., 30), he ultimately decides to limit membership to a small subgroup of mammals. In fact, he restricts the term "animal" to this particular class of entities. "Unless otherwise indicated," Regan reports, "the word *animal* will be used to refer to mentally normal mammals of a year or more" (ibid., 78).

Regan, it should be noted, is not alone in this exclusive decision. It has also been advanced, albeit for very different reasons, by John Searle (1997, 5), who rather confidently operationalizes consciousness "as an inner, first-person, qualitative phenomenon." Following this definition, Searle draws the following conclusion: "Humans and higher animals are obviously conscious, but we do not know how far down the phylogenetic scale consciousness extends. Are fleas conscious, for example? At the present state of neurobiological knowledge, it is probably not useful to worry about such questions" (ibid.). Like Regan, Searle also recognizes the obvious problem of drawing the line of demarcation but then immediately excuses himself from giving it any further consideration. Although justified either in terms of "economy of expression," as Regan (1983, 83) proposes, or Searle's appeal to utility, this decision is no less prejudicial and exclusive than the one that had been instituted by Descartes. Regan's *The Case for Animal Rights*, therefore, simply replaces the Cartesian bias against all nonhuman animals with a more finely tuned prejudice against some animals. Although extending the field of morality by including some nonhuman animals, these efforts do so by reproducing the same exclusive decision, one that effectively marginalizes many, if not most, animals.

On the other hand, that other figure of excluded otherness, the machine, also appears to have made successful claims on consciousness. Although the instrumentalist viewpoint precludes ascribing anything approaching consciousness to technological artifacts like computers, robots, or other mechanisms, the fact is machines have, for quite some time, disputed this decision in both science fiction and science fact. The issue is, for example, directly addressed in the course of a fictional BBC television documentary that is included (as a kind of Shakespearean "play within a play") in *2001: A Space Odyssey.*

BBC Interviewer: HAL, despite your enormous intellect, are you ever frustrated by your dependence on people to carry out actions?
HAL: Not in the slightest bit. I enjoy working with people. I have a stimulating relationship with Dr. Poole and Dr. Bowman. My mission responsibilities range over the entire operation of the ship, so I am constantly occupied. I am putting myself to the fullest possible use, which is all I think that any conscious entity can ever hope to do.

When directly questioned, HAL not only responds in a way that appears to be conscious and self-aware but also refers to himself as a thinking "conscious entity." Whether HAL actually is conscious, as opposed to being merely designed to appear that way, is a question that is, as far as the human crew is concerned, ultimately undecidable.

BBC Interviewer: In talking to the computer, one gets the sense that he is capable of emotional responses, for example, when I asked him about his abilities, I sensed a certain pride in his answer about his accuracy and perfection. Do you believe that HAL has genuine emotions?

Dave: Well, he acts like he has genuine emotions. Um, of course he's programmed that way to make it easier for us to talk to him, but as to whether or not he has real feelings is something I don't think anyone can truthfully answer.

Although the HAL 9000 computer is a fictional character, its features and operations are based on, derived from, and express the very real objectives of AI research, at least as they had been understood and developed in the latter half of the twentieth century. The achievement of human-level intelligence and conscious behavior, what is often called following John Searle's (1997, 9) terminology "strong AI," was considered a suitable and attainable goal from the very beginning of the discipline as set out at the Dartmouth conference in 1956. And this objective, despite the persuasive efforts of critics, like Joseph Weizenbaum, Hubert Dreyfus, John Searle, and Roger Penrose, as well as recognized setbacks in research progress, is still the anticipated outcome predicted by such notable figures as Hans Moravec, Ray Kurzweil, and Marvin Minsky, who it will be recalled consulted with Stanley Kubrick and his production team on the design of HAL. "The ultimate goal of machine cognition research," Haikonen (2007, 185) writes, "is to develop autonomous machines, robots and systems that know and understand what they are doing, and are able to plan, adjust and optimize their behavior in relation to their given tasks in changing environments. A system that succeeds here will most probably appear as a conscious entity." And these "conscious machines" are, at least in the opinion of experts, no longer some distant possibility. "In May of 2001," Owen Holland (2003, 1) reports, "the Swartz Foundation sponsored a workshop called 'Can a machine be conscious?' at the Banbury Center in Long Island. Around twenty psychologists, computer scientists, philosophers, physicists, neuroscientists, engineers, and industrialists spent three days in a mixture of short presentations and long and lively discussions. At the end,

Christof Koch, the chair, asked for a show of hands to indicate who would now answer 'Yes' to the question forming the workshop theme. To everyone's astonishment, all hands but one were raised."

Despite the fact that human-level consciousness is something that is still located just over the horizon of possibility—perhaps even endlessly deferred and protected as a kind of Platonic ideal—there are working prototypes and practical research endeavors that provide persuasive and convincing evidence of machines that have been able to achieve some aspect of what is considered "consciousness." One promising approach has been advanced by Raul Arrabales, Agapito Ledezma, and Araceli Sanchis (2009) as part of the ConsScale project. ConsScale is a proposed consciousness metric derived from observations of biological systems and intended to be used both to evaluate achievement in machine consciousness and to direct future design efforts. "We believe," Arrabales, Ledezma, and Sanchis (2009, 4) argue, "that defining a scale for artificial consciousness is not only valuable as a tool for MC [machine consciousness] implementations comparative study, but also for establishing a possible engineering roadmap to be followed in the quest for conscious machines." As proof of concept, the authors apply their scale to the evaluation of three software bots designed and deployed within "an experimental environment based on the first-person shooter video game Unreal Tournament 3" (ibid., 6). The results of the study demonstrate that these very rudimentary artificial entities exhibited some of the benchmark qualifications for consciousness as defined and characterized by the ConsScale metric.

Similarly, Stan Franklin (2003, 47) introduces a software agent he calls IDA that is "functionally conscious" insofar as "IDA perceives, remembers, deliberates, negotiates, and selects actions." "All of this together," Franklin concludes, "makes a strong case, in my view, for functional consciousness" (ibid., 63). However, what permits IDA to be characterized in this fashion depends, as Franklin is well aware, on the way consciousness comes to be defined and operationalized. But even if we employ the less restricted and more general definition of what David Chalmers (1996) calls "phenomenal consciousness," the outcome is equivocal at best. "What about phenomenal consciousness?" Franklin (2003, 63) asks. "Can we claim it for IDA? Is she *really* a conscious artifact? I can see no convincing arguments for such a claim. . . . On the other hand, I can see no convincing arguments against a claim for phenomenal consciousness in IDA."

This undecidability, resulting from actual experience with working prototypes, is further complicated by theoretical inconsistencies in the arguments often made in opposition to machine consciousness. Hilary Putnam identifies the source of the trouble in his seminal article "Robots: Machines or Artificially Created Life?":

All these arguments suffer from one unnoticed and absolutely crippling defect. They rely on just two facts about robots: that they are artifacts and that they are deterministic systems of a physical kind, whose behavior (including the "intelligent" aspects) has been preselected and designed by an artificer. But it is purely contingent that these two properties are *not* properties of human beings. Thus, if we should one day discover that *we* are artifacts and that our every utterance was anticipated by our superintelligent creators (with a small "c"), it would follow, if these arguments were sound, that *we* are not *conscious*! At the same time, as just noted, these two properties are not properties of all imaginable robots. Thus these two arguments fail in two directions: they might "show" that *people are not* conscious—because people might be the wrong sort of robot—while simultaneously failing to show that some robots are not conscious. (Putnam 1964, 680)

According to Putnam, the standard instrumentalist conceptualization, which assumes that robots and other machines are mere instruments or artifacts, the behavior of which is preselected and determined by a human designer or programmer, is something that, if rigorously applied, would fail in two ways. On the one hand, it could lead to the conclusion that humans are not conscious, insofar as an individual human being is created by his or her parents and determined, in both form and function, by instructions contained in genetic code. Captain Picard, Data's advocate in the *Star Trek: The Next Generation* episode "Measure of a Man," draws a similar conclusion: "Commander Riker has dramatically demonstrated to this court that Lieutenant Commander Data is a machine. Do we deny that? No, because it is not relevant—we too are machines, just machines of a different type. Commander Riker has also reminded us that Lieutenant Commander Data was created by a human; do we deny that? No. Again it is not relevant. Children are created from the 'building blocks' of their parents' DNA." On the other hand, this mechanistic determination fails to take into account all possible kinds of mechanisms, especially learning automata. Machines that are designed for and are able to learn do not just do what was preprogrammed but often come up with unique solutions that can even surprise their programmers. According to Putnam, then, it would not be possible to prove, with anything approaching certainty, that these machines were *not* conscious. Like astronaut Dave Bowman, the best anyone can do in these circumstances is to admit that the question regarding machine consciousness cannot be truthfully and definitively answered.

The main problem in all of this is not whether animals and machines are conscious or not. This will most likely remain a contentious issue, and each side of the debate will obviously continue to heap up both practical examples and theoretical arguments to support its own position. The real problem, the one that underlies this debate and regulates its entire operations, is the fact that this discussion proceeds and persists with a rather flexible and not entirely consistent or coherent characterization of consciousness. As Rodney Brooks (2002, 194) admits, "we have no real operational definition of consciousness," and for that reason, "we are completely prescientific at this point about what consciousness is." Relying, for example, on "folk psychology," as Haikonen (2007, 2) points out, "is not science. Thus it is not able to determine whether the above phenomena were caused by consciousness or whether consciousness is the collection of these phenomena or whether these phenomena were even real or having anything to do with consciousness at all. Unfortunately philosophy, while having done much more, has not done much better." Although consciousness, as Anne Foerst remarks, is the secular and supposedly more "scientific" replacement for the occultish "soul" (Benford and Malartre 2007, 162), it turns out to be just as much an occult property.

The problem, then, is that consciousness, although crucial for deciding who is and who is not a person, is itself a term that is ultimately undecided and considerably equivocal. "The term," as Max Velmans (2000, 5) points out, "means many different things to many different people, and no universally agreed core meaning exists." And this variability often has an adverse effect on research endeavors. "Intuitive definitions of consciousness," as Arrabales, Ledezma, and Sanchis (2009, 1) recognize, "generally involve perception, emotions, attention, self-recognition, theory of mind, volition, etc. Due to this compositional definition of the term consciousness it is usually difficult to define both what is exactly a conscious being and how consciousness could be implemented in artificial machines." Consequently, as Dennett (1998, 149–150) concludes, "consciousness appears to be the last bastion of occult properties, epiphenomena, immeasurable subjective states" comprising a kind of impenetrable "black box." In fact, if there is any general agreement among philosophers, psychologists, cognitive scientists, neurobiologists, AI researchers, and robotics engineers regarding consciousness, it is that there is little or no agreement when it comes to defining and characterizing the concept. And to make matters worse, the problem is not just with the lack of a basic definition; the problem may itself already be a problem. "Not only is there no consensus on what the term *consciousness* denotes," Güven Güzeldere (1997, 7) writes, "but neither is it immediately clear if there actually is a single, well-defined '*the* problem of consciousness' within disciplinary (let alone across disciplinary) boundaries. Perhaps the trouble lies not so much in the ill definition of the question, but in the fact that what passes under the term consciousness as an all too familiar, single, unified notion may be a tangled amalgam of several different concepts, each inflicted with its own separate problems."

1.4.2 Turing Tests and Other Demonstrations

Defining one or more personal properties, like consciousness, is only half the problem. There is also the difficulty of discerning the presence of one or more of the properties in a particular entity. That is, even if we could agree on some definition of consciousness, for example, we would still need some way to detect and prove that someone or something, human, animal, or otherwise, actually possessed it. This is, of course, a variant of "the problem of other minds" that has been a staple of the philosophy of mind from its inception. "How does one determine," as Paul Churchland (1999, 67) characterizes it, "whether something other than oneself—an alien creature, a sophisticated robot, a socially active computer, or even another human—is really a thinking, feeling, conscious being; rather than, for example, an unconscious automaton whose behavior arises from something other than genuine mental states?" Or to put it in the more skeptical language employed by David Levy (2009, 211), "how would we know whether an allegedly Artificial Conscious robot really was conscious, rather than just behaving-as-if-it-were-conscious?" And this difficulty, as Gordana Dodig-Crnkovic and Daniel Persson (2008, 3) explain, is rooted in the undeniable fact that "we have no access to the inner workings of human minds—much less than we have access to the inner workings of a computing system." In effect, we cannot, as Donna Haraway (2008, 226) puts it, climb into the heads of others "to get the full story from the inside." Consequently, attempts to resolve or at least respond to this problem almost always involve some kind of behavioral observation, demonstration, or empirical testing. "To put this another way," Roger Schank (1990, 5) concludes, "we really cannot examine the insides of an intelligent entity in such a way as to establish what it actually knows. Our only choice is to ask and observe."

This was, for instance, a crucial component of the petition filed on behalf of the chimpanzee Matthew Hiasl Pan and adjudicated by the Austrian courts. "Within a behavioural enrichment project," Balluch and Theuer explain in their review of the case,

Matthew has passed a mirror self-recognition test, he shows tool use, plays with human caretakers, watches TV and draws pictures. Matthew can understand if caretakers want to lure him into doing something, and then decides whether this is in his interest or not. He can pretend to feel or want something when actually he has other intentions thus showing that he deliberately hides his real intentions in order to achieve his aims. Those humans close to him, who know him best, clearly support the proposition that he has a theory of mind and does understand intentional states in other persons. (Balluch and Theuer 2007, 1)

To justify extension of the term "person" to a chimpanzee, Balluch and Theuer cite a number of psychological and behavioral tests that are designed for and recognized by a particular community of researchers as providing credible evidence that this nonhuman animal does in fact possess one or more of the necessary personal properties. A similar kind of demonstration would obviously be necessary to advance a related claim for an intelligent machine or robot, and the default demonstration remains the Turing test, or what Alan Turing (1999), its namesake, initially called "the imitation game." If a machine, Turing hypothesized, becomes capable of successfully simulating a human being in communicative exchanges with a human interlocutor, then that machine would need to be considered "intelligent." Although initially introduced for and limited to deciding the question of machine intelligence, the test has been extended to the question concerning personhood.

This is, for example, the situation in Leiber's fictional dialogue, where both sides of the debate mobilize versions of the Turing test to support their positions. On the one side, advocates for including a computer, like the fictional AL, in the community of persons employ the test as way to demonstrate machine consciousness. "I submit," the complainant in the hypothetical hearing argues, "that current computers, AL in particular, can play a winning game of imitation. AL can pass the Turing test. Mentally speaking, AL can do what a human being can do. Indeed, the human crew of Finland Station interacted with AL as if AL were a kindly, patient, confidential, and reliable uncle figure" (Leiber 1985, 26). According to this line of argumentation, the space station's central computer should be considered a person, because it behaves and was treated by the human crew as if it were another human person.

On the other side, it is argued that what AL and similarly constructed machines actually do is simply manipulate symbols, taking input and spitting out preprogrammed output, much like Joseph Weizenbaum's (1976) ELIZA chatter-bot program or John Searle's (1980) Chinese room thought experiment. And the respondent in the fictional hearing mobilizes both examples, in order to argue that what happens inside AL is nothing more than "an endless manipulation of symbols" (Leiber 1985, 30) that is effectively mindless, unconscious, and without intelligence. "How can this moving about mean anything, or mount up to a person who has meaningful thoughts and emotions, and a sense of personhood? Indeed, maybe all Turing's suggestion amounts to is that a computer is a generalized symbol-manipulating device, ultimately a fantastically complicated network of off-on switches, not something you can think of as a person, as something to care about?" (ibid.). Consequently (to mobilize terminology that appears to saturate this debate), such mechanisms are merely capable of *reacting* to input but are not actually able to *respond* or act responsibly.

Deploying the Turing test in this fashion is not limited to this fictional account but has also had considerable traction in the current debates about personhood, consciousness, and ethics. David Levy, for instance, suggests that the question of machine consciousness, which continues to be a fundamental component of roboethics, should be approached in the same way that Turing approached the question of intelligence: "To summarize and paraphrase Turing, if a machine exhibits behavior that is normally a product of human intelligence, imagination for example, or by recognizing sights and scenes and music and literary style, then we should accept that that machine is intelligent. Similarly, I argue that if a machine exhibits behavior of a type normally regarded as a product of human consciousness (whatever consciousness might be), then we should accept that that machine has consciousness" (Levy 2009, 211). This approach to testing other kinds of entities, however, also has important precursors, and we find a version of it administered to both animals and machines in the course of Descartes's *Discourse on Method*. In fact, it could be argued that the Cartesian test or "game of imitation" comprises the general prototype and model for all subsequent kinds of testing, whether designed for and administered to animals or machines. Although not using its formalized language, Descartes begins from the defining condition of the other minds problem. He indicates how, if one were following a strict method of observational analysis, that he or she would be unable to decide with any certainty whether what appears as other men on the street are in fact real men and not cleverly designed automatons.

This fundamental doubt about everything and everyone else beyond oneself, or what is often called solipsism, is not something limited to the Cartesian method. It is shared by contemporary researchers working in a number of different fields (e.g., philosophy of mind, psychology, computer-mediated communication) and it has been a perennial favorite in science fiction. "Epistemological debates about the existence and knowability of 'other minds,'" Judith Donath (2001, 298) argues in a consideration of computer-mediated communication and software bots, "often poses a skeptical view hypothesizing that the other person may actually be a robot or other nonconscious being. The mediated computational environment makes this a very real possibility." Likewise, Auguste Villiers de l'Isle-Adam's *L'Eve future* (1891) or *Tomorrow's Eve* (2001), the symbolist science fiction novel that initially popularized the term "android" (*andreide*), gets a good deal of narrative mileage out of the potential confusion between real people and the artificial imitation of a human being (Villiers de l'Isle-Adam 2001, 61). According to Carol de Dobay Rifelj (1992, 30), "the problem of other minds is often posed as a question whether the other knows anything at all, whether other people might not be just robots. Villiers de l'Isle-Adam's *Tomorrow's Eve* raises it in a very concrete way, because it recounts the construction of an automaton that is to take the place of a real woman. Whether the man for whom 'she' is constructed can accept her as a person is crucial for the novel, which necessarily broaches the issue of consciousness and human identity." The substitutability of real and artificial women is also a crucial narrative component of Fritz Lang's *Metropolis* (1927), in which Rotwang's highly sexualized robot takes the place of the rather modest Maria in order to foment rebellion in the worker's city. The fact that these prototypical literary and cinematic androids are gendered female is no accident. This is because, within the Western tradition at least, there has been serious (albeit terribly misguided) doubt as to whether women actually possessed rational minds or not. It should also be noted that the names of these artificial females are historically significant. Eve, of course, refers to the first woman who leads Adam into sin, and Maria references the virgin mother of Jesus Christ.

Despite potential confusion, there are, at least according to Descartes (1988, 44), two "very certain means of recognizing" that these artificial figures are in fact machines and not real men (or women):

The first is that they could never use words, or put together other signs, as we do in order to declare our thoughts to others. For we can certainly conceive of a machine so constructed that it utters words, and even utters words which correspond to bodily actions causing a change in its organs. But it is not conceivable that such a machine should produce different arrangements of words so as to give an appropriately meaningful answer to whatever is said in its presence, as the dullest of men can do. Secondly, even though such machines might do some things as well as we do them, or perhaps even better, they would inevitably fail in others, which would reveal that they were acting not through understanding but only from the disposition of their organs. For whereas reason is a universal instrument which can be used in all kinds of situations, these organs need some particular disposition for each particular action; hence it is for all practical purposes impossible for a machine to have enough different organs to make it act in all the contingencies of life in the way in which our reason makes us act. (Ibid., 44–45)

For Descartes, what distinguishes a human-looking machine from an actual human being is the fact that the former obviously and unquestionably lacks both language and reason. These two components are significant because they constitute the two concepts that typically are employed to translate the Greek term λόγος. In fact, the human being, beginning with the scholastic philosophers of the medieval period and continuing through the innovations of the modern era, had been defined as *animal rationale*, a living thing having reason. This characterization, as Martin Heidegger (1962, 47) points out, is the Latin translation and interpretation of the Greek ζῳον λόγον ἐχον. Although λόγος has been routinely translated as "*ratio*," "rationality," or "reason," Heidegger demonstrates that the word literally indicated word, language, and discourse. The human entity, on this account, does not just possess reason and language as faculties but is defined by this very capacity. Consequently, λόγος—reason and/or language—is definitive of the human and for this reason has been determined, as Descartes demonstrates, to be something restricted to the human subject. In other words, the automaton, although capable of having the external shape and appearance of a man, is absolutely unable to "produce different arrangements of words so as to give an appropriately meaningful answer to whatever is said in its presence" (Descartes 1988, 44). As Derrida (2008, 81) points out, it may be able to *react*, but it cannot *respond*. Furthermore, it does not possess nor is it capable of simulating the faculty of reason, which is, according to Descartes's explanation, the universal instrument that directs all human endeavor.

Because the animal and machine share a common ontological status, what is often called the Cartesian *bête-machine*, Descartes (1988) immediately employs this particular association to describe and differentiate the animal.

Now in just these two ways we can also know the difference between man and beast. For it is quite remarkable that there are no men so dull-witted or stupid—and this includes even madmen—that they are incapable of arranging various words together and forming an utterance from them in order to make their thoughts understood; whereas there is no other animal, however perfect and well-endowed it may be, that can do the like. . . . This shows not merely that the beasts have less reason than men, but that they have no reason at all. For it patently requires very little reason to be able to speak; and since as much inequality can be observed among the animals of a given species as among human beings, and some animals are more easily trained than others, it would be incredible that a superior specimen of the monkey or parrot species should not be able to speak as well as the stupidest child—or at least as well as a child with a defective brain—if their souls were not completely different in nature from ours. (Descartes 1988, 45)

According to this Cartesian argument, the animal and the machine are similar insofar as both lack the ability to speak and, on the evidence of this deficiency, also do not possess the faculty of reason. Unlike human beings, who, despite various inequalities in actual capabilities, can speak and do possess reason, the animal and machine remain essentially speechless and irrational. In short, neither participates in λόγος. Consequently, this Cartesian demonstration organizes the animal and machine under one form of alterity. Both are the *same* insofar as both are completely *other* than human. In fact, according to this line of argument, there can be no reliable way to distinguish between a machine and an animal. Although a real human being is clearly distinguishable from a human-looking automaton, there is, on Descartes's account, no way to differentiate an animal automaton from a real animal. If we were confronted, Descartes argues, with a machine that mimics the appearance of a monkey or any other creature that lacks reason, there would be no means by which to distinguish this mechanism from the actual animal it simulates (Descartes 1988, 44).

Descartes's insights, which in the early seventeenth century may have been able to be written off as theoretical speculation, have been prototyped in both science fact and science fiction. Already in 1738, for example, the argument concerning the *bête-machine* was practically demonstrated when Jacques de Vaucanson exhibited a mechanical duck, which reportedly was indistinguishable from a real duck. More recent demonstrations have been staged in Rodney Brooks's lab, where robotic insectlike creatures, with names like Genghis, Attila, and Hannibal, appear to move and react in ways that are virtually indistinguishable from a real animal. "When it was switched on," Brooks (2002, 46) writes concerning Genghis, "it came to life! It had a wasp like personality: mindless determination. But it had a personality. It chased and scrambled according to its will, not to the whim of a human controller. It acted like a creature, and to me and others who saw it, it felt like a creature. It was an artificial creature."

A similar situation, one that capitalizes on every aspect of the Cartesian text, is dramatically illustrated in Philip K. Dick's *Do Androids Dream of Electric Sheep?* (1982), the science fiction novel that provided the raw material for the film *Blade Runner*. In Dick's post-apocalyptic narrative, nonhuman animals are all but extinct. Because of this, there is great social capital involved in owning and caring for an animal. However, because of their scarcity, possessing an actual animal is prohibitively expensive. Consequently, many people find themselves substituting and tending to animal automatons, like the electric sheep of the title. For most individuals, there is virtually no way to distinguish the electric sheep from a real one. Like Vaucanson's duck, both kinds of sheep eat, defecate, and bleat. In fact, so perfect is the illusion that when an electric animal breaks down, it is programmed to simulate the pathology of illness, and the repair shop, which is complicit in the deception, operates under the pretense of a veterinary clinic. At the same time, this desolate and depopulated world is also inhabited by human automatons or androids. Whereas the confusion between the animal and machine is both acceptable and propitious, the same cannot be said of the human-looking automaton. The "replicants," which are what these androids are called, must be rooted out, positively identified, and, in a carefully selected euphemism, "retired." Although there is no practical way to differentiate the animal from the machine other than destructive analysis or dissection, there is, according to Dick's narrative, a reliable way to differentiate an automaton from an actual human being. And the evaluation involves conversational interaction. The suspected android is asked a series of questions and, depending upon his or her response in dialogue with the examiner, will, in a kind of perverse Turing test, eventually betray its artificial nature.

For Descartes, as for much of modern European-influenced thought, the distinguishing characteristic that had allowed one to divide the human being from its others, the animal and machine, is λόγος. In fact, it seems that there is a closer affinity between the animal and machine owing to a common lack of λόγος than there is between the human and animal based on the common possession of ζωον—life. In other words, it appears that discourse and reason trump life, when it comes to dividing us from them. This strategy, however, is no longer, and perhaps never really was, entirely successful. In 1967, for example, Joseph Weizenbaum, already demonstrated a very simple program that simulated conversational exchange with a human interlocutor. ELIZA, the first chatter-bot, was able to converse with human users by producing, to redeploy the words of Descartes (1988, 44), "different arrangements of words so as to give an appropriately meaningful answer to whatever is said in its presence." Because of experience with machines like ELIZA and more sophisticated chatter-bots now deployed in virtual environments and over the Internet, the boundary between the human animal and the machine has become increasingly difficult to distinguish and defend. Similar discoveries have been reported with nonhuman animals. If machines are now capable of some form of discursive communication, then it should be no surprise that animals have also been found to display similar capabilities. Various experiments with primates, like those undertaken by Sue Savage-Rumbaugh and company (1998), have confirmed the presence of sophisticated linguistic abilities once thought to be the exclusive possession of human beings. According to Carey Wolfe (2003a, xi), "a veritable explosion of work in areas such as cognitive ethology and field ecology has called into question our ability to use the old saws of anthropocentrism (language, tool use, the inheritance of cultural behaviors, and so on) to separate ourselves once and for all from the animals, as experiments in language and cognition with great apes and marine mammals, and field studies of extremely complex social and cultural behaviors in wild animals such as apes, wolves, and elephants, have more or less permanently eroded the tidy divisions between human and nonhuman."

The problem gets even more complicated if we consider it from the perspective of reason or rationality. Although considered at one time to be the defining characteristic of the human being, reason can no longer be, and perhaps never really was, an exclusively human faculty. *Ratio*, as Heidegger (1996, 129) reminds us, is a word that was originally adopted from Roman commercial discourse around the time of Cicero and identified, prior to indicating anything like "thought" or "cognition" in general, the specific operations of accounting, reckoning, and calculation. Gottfried Wilhelm von Leibniz, who was critical of the Cartesian innovations, illustrated this fundamental connection in his planned *De arte combinatoria*, a project that endeavored "to create a general method in which all truths of reason would be reduced to a kind of calculation" (Haaparanta 2009, 135). In fact, Leibniz's objective, one that he pursued throughout his professional career but never actually completed, was to create a rational calculus that would resolve all philosophical problems and controversy through mechanical calculation rather than by way of impassioned debate and discussion. Currently computers not only outperform human operators in mathematical operations and the proof of complex theorems but also translate between human languages, beat grand-master champions at chess, and play improvisational jazz. As Brooks concludes, reason no longer appears to be the defining barrier we once thought it was. "Just as we are perfectly willing to say that an airplane can fly, most people today, including artificial intelligence researchers, are willing to say that computers, given the right set of software and the right problem domain, *can* reason about facts, *can* make decisions, and *can* have goals" (Brooks 2002, 170).

Not only are machines able to emulate and in some instances even surpass human reason, but some theorists now argue that machines, and not human beings, are the only rational agents. Such an argument, pitched in distinctly moral terms, is advanced by Joseph Emile Nadeau in his posthumously published essay, "Only Androids Can Be Ethical." "Responsibility and culpability," Nadeau (2006, 245) writes, "require action caused by a free will, and such action suffices for an entity to be subject to ethical assessment to be ethical or unethical. An action is caused by free will if and only if it is caused by reasons. Human actions are not, save possibly very rarely, caused by reasons. The actions of an android built upon a theorem prover or neural network or some combination of these could be caused by reasons. Hence an android, but not a human, could be ethical." Moral reasoning requires, whether one follows Kant's deontological ethics or Bentham's utilitarian "moral calculus," rational decision making. Humans, according to Nadeau's argument, are unfortunately not very rational, allowing for decisions to be influenced by emotional attachments and unsubstantiated judgments. Machines, however, can be programmed with perfect and infallible logical processing. Therefore, Nadeau concludes, only machines can be fully rational; and if rationality is the basic requirement for moral decision making, only a machine could ever be considered a legitimate moral agent. For Nadeau, the main issue is not whether and on what grounds machines might be admitted to the population of moral persons, but whether human beings should qualify in the first place.

The real issue in this debate, however, is not proving whether an animal or machine does or does not possess the requisite person-making qualities by way of argumentation, demonstration, or testing. The problem is more fundamental. As both Dennett (1998) and Derrida (2008) point out, albeit in very different contexts, the real problem is the unfounded inference that both sides of the debate endorse and enact—the leap from some externally observable phenomenon to a presumption (whether negative or positive) concerning internal operations, which are then *(presup)posited*, to use Žižek's (2008a, 209) neologism, as the original cause and referent of what is externally available. This insight, in fact, is rooted in and derived from the critical work of Immanuel Kant. In the *Critique of Pure Reason*, Kant famously argued that a thing is to be taken in a twofold sense, the thing as it appears to us and the thing as it is in itself (*das Ding an sich*). Kant's point is that one cannot make inferences about the latter from the experiences of the former without engaging in wild and unfounded speculation. Consequently (and extending this Kantian insight in a direction that Kant would not necessarily endorse), whether another human being, or any other thing, really does or does not possess the capabilities that it appears to exhibit is something that is ultimately undecidable. "There is," as Dennett (1998, 172) concludes, "no proving [or disproving] that something that seems to have an inner life does in fact have one—if by 'proving' we understand, as we often do, the evincing of evidence that can be seen to establish by principles already agreed upon that something is the case." Although philosophers, psychologists, and neuroscientists throw an incredible amount of argumentative and experimental force at this "other minds" problem, it is not able to be resolved in any way approaching what would pass for good empirical science. In the end, not only are these tests unable to demonstrate with any credible results whether animals and machines are in fact conscious and therefore legitimate persons (or not), we are left doubting whether we can even say the same for other human beings. As Kurzweil (2005, 378) candidly concludes, "we assume other humans are conscious, but even that is an assumption," because "we cannot resolve issues of consciousness entirely through objective measurement and analysis (science)" (ibid., 380).

1.5 Personal Problems and Alternatives

If anything is certain from this consideration of the concept, it is that the term "person," the attendant "person-making qualities," and the different approaches to detection and demonstration have been equivocal at best. The concept obviously carries a good deal of metaphysical and ethical weight, but what it consists in remains ultimately unresolved and endlessly debatable. For some, like David DeGrazia, this equivocation is not necessarily a problem. It is both standard procedure and a considerable advantage:

I suggest that personhood is associated with a cluster of properties without being precisely definable in terms of any specific subset: autonomy, rationality, self-awareness, linguistic competence, sociability, the capacity for action, and moral agency. One doesn't need all these traits, however specified, to be a person, as demonstrated by nonautonomous persons. Nor is it sufficient to have just one of them, as suggested by the fact that a vast range of animals are capable of intentional action. Rather, a person is someone who has enough of these properties. Moreover the concept is fairly vague in that we cannot draw a precise, nonarbitrary line that specifies what counts as enough. Like many or most concepts, personhood has blurred boundaries. Still person means something, permitting us to identify paradigm persons and, beyond these easy cases, other individuals who are sufficiently similar to warrant inclusion under the concept. (DeGrazia 2006, 42–43)

For DeGrazia, the absence of a precise definition and lack of a stable characterization for the term "person" is not necessarily a deal breaker. Not only are other important concepts beset by similar difficulties, but it is, DeGrazia argues, precisely this lack of precision that allows one to make a case for including others. In other words, tolerating some slippage and flexibility in the definition of the concept allows for "personhood" to remain suitably open and responsive to other, previously excluded groups and individuals. At the same time, however, this conceptual flexibility should be cause for concern insofar as it renders important decisions about moral status—especially decisions concerning who or what is included and who or what is not—capricious, potentially inconsistent, and ultimately relative. And this is not just a metaphysical puzzle; it has significant moral consequences. "Our assumption that an entity is a person," Dennett (1998, 285) writes, "is shaken precisely in those cases where it really matters: when wrong has been done and the question of responsibility arises. For in these cases the grounds for saying that the person is culpable (the evidence that he did wrong, was aware he was doing wrong, and did wrong of his own free will) are in themselves grounds for doubting that it is a person we are dealing with at all. And if it is asked what could settle our doubts, the answer is: nothing."

To complicate matters, all these things are referred to and ultimately evaluated and decided by an interested party. "The debate about whether computer systems can ever be 'moral agents' is a debate among humans about what they will make of computational artifacts that are currently being developed" (Johnson and Miller 2008, 132). It is, then, human beings who decide whether or not to extend moral agency to machines, and this decision itself has ethical motivations and consequences. In other words, one of the parties who stand to benefit or lose from these determinations is in the position of adjudicating the matter. Human beings, those entities who are already considered to be persons, not only get to formulate the membership criteria of personhood but also nominate themselves as the deciding factor. In this way, "the ethical landscape," as Lucas Introna (2003, 5) describes it, "is already colonised by humans. . . . It is us humans that are making the decisions about the validity, or not, of any criteria or category for establishing ethical significance. . . . Are not all our often suggested criteria such as originality, uniqueness, sentience, rationality, autonomy, and so forth, not somehow always already based on that which we by necessity comply with?" This means, in effect, that "man is the measure of all things" in these matters. Human beings not only get to define the standard qualifying criteria, which are often based on and derived from their own abilities and experiences, but also nominate themselves both judge and jury for all claims on personhood made by or on behalf of others. Consequently, instead of providing an objective and equitable orientation for ethics, the concept *person* risks reinstalling human exceptionalism under a different name. Although the concept *person* appears to open up moral thought to previously excluded others, it does so on exclusively human terms and in a way that is anything but altruistic.

1.5.1 Rethinking Moral Agency

Contending with the question of moral agency as it is currently defined appears to lead into that kind of intellectual cul-de-sac or stalemate that Hegel (1969, 137) called a "bad infinite." "The debate," as Deborah Johnson (2006, 195) argues, "seems to be framed in a way that locks the interlocutors into claiming either that computers are moral agents or that computers are not moral." Formulated in this fashion the two sides are situated as dialectical opposites with the one negating whatever is advanced or argued by the other. As long as the debate continues to be articulated in this manner it seems that very little will change. To make some headway in this matter, Johnson suggests altering our perspective and reconfiguring the terms of the debate. "To deny that computer systems are moral agents is not the same as denying that computers have moral importance or moral character; and to claim that computer systems are moral is not necessarily the same as claiming that they are moral agents. The interlocutors neglect important territory when the debate is framed in this way. In arguing that computer systems are moral entities but are not moral agents, I hope to reframe the discussion of the moral character of computers" (ibid.).

According to Johnson, the way the debate is currently defined creates and perpetuates a false dichotomy. It misses the fact that the two seemingly opposed sides are not necessarily mutually exclusive. That is, she contends, it is possible both to reserve and to protect the concept of moral agency by restricting it from computers, while also recognizing that machines are ethically important or have some legitimate claim on moral behavior:

My argument is that computer systems do not and cannot meet one of the key requirements of the traditional account of moral agency. Computer systems do not have mental states and even if states of computers could be construed as mental states, computer systems do not have intendings to act arising from their freedom.

Thus, computer systems are not and can never be (autonomous, independent) moral agents. On the other hand, I have argued that computer systems have intentionality, and because of this, they should not be dismissed from the realm of morality in the same way that natural objects are dismissed. (Ibid., 204)

In this way, Johnson argues for making fine distinctions in the matter of moral action and "intentionality." Unlike human beings, computers do not possess mental states, nor do they give evidence of intendings to act arising from their freedom. But unlike natural objects, for example, Kaspar Hauser's apples or Descartes's animals, computers do not simply "behave from necessity" (ibid.). They have intentionality, "the intentionality put into them by the intentional acts of their designers" (ibid., 201). Reframing the debate in this fashion, then, allows Johnson to consider the computer as an important player in ethical matters but not a fully constituted moral agent. "Computer systems are components in moral action," Johnson (ibid., 204) concludes. "When humans act with artifacts, their actions are constituted by their own intentionality and efficacy as well as the intentionality and efficacy of the artifact which in turn has been constituted by the intentionality and efficacy of the artifact designer. All three—designer, artifact, and users—should be the focus of moral evaluation."

Although Johnson's (ibid., 202) "triad of intentionality" is more complex than the standard instrumentalist position, it still proceeds from and protects a fundamental investment in human exceptionalism. Despite considerable promise to reframe the debate, Johnson's new paradigm does not look much different from the one it was designed to replace. Human beings are still and without question the only legitimate moral agents. Computers might complicate how human intentionality is distributed and organized, but they do not alter the fundamental "fact" that human beings are and remain the only moral agents. A more radical reformulation proceeds from attempts to redefine the terms of agency so as to be more inclusive. This is possible to the extent that moral agency is, to begin with, somewhat flexible and indeterminate. "There are," Paul Shapiro argues,

many ways of defining moral agency and the choice of a definition is a crucial factor in whether moral agency proves to be limited to humans. Philosophers like Pluhar set the standard for moral agency at a relatively high level: the capability to understand and act on moral principles. In order to meet this standard, it seems necessary for a being to possess linguistic capacities beyond those presently ascribed to any other species (with the possible exception of some language-trained animals). However, a lower standard for moral agency can also be selected: the capacity for virtuous behavior. If this lower standard is accepted, there can be little doubt that many other animals are moral agents to some degree. (Shapiro 2006, 358)

As Shapiro recognizes, who is and who is not included in the community of moral agents is entirely dependent upon how "moral agent" is defined and characterized, and changes in the definition can provide for changes in the population, having the effect of either including or excluding others. Depending on where and how the line is drawn, traditionally excluded figures, like animals and machines, will either be situated outside the circle or admitted into the club. Consequently, a lot is riding on how agency is characterized, who gets to provide the characterization, and how that configuration is positioned and justified.

John P. Sullins (2006), for instance, recognizes that as long as moral agency is associated with personhood, machines will most likely never achieve the status of being a moral subject. They will continue to be mere instruments used, more or less effectively, by human persons for humanly defined ends. Sullins, therefore, endeavors to distinguish the two terms. That is, he affirms "that the robots of today are certainly not persons" but argues "that personhood is not required for moral agency" (Sullins 2006, 26). His demonstration of this begins by outlining the four "philosophical views on the moral agency of robots." The first is exemplified by Dennett (1998), who, according to Sullins's reading of the HAL essay, argues that "robots are not now moral agents but might become them in the future" (Sullins 2006, 26). This position holds open the possibility of machine moral agency but postpones any definitive decision on the matter. The second is exemplified by the work of Selmer Bringsjord (2008), who argues, following the precedent of instrumentalism and in direct opposition to the former viewpoint, that computers and robots "will never do anything they are not programmed to perform," and as a result "are incapable of becoming moral agents now or in the future" (Sullins 2006, 26). A third, albeit much less popular, alternative can be found in Joseph Emile Nadeau's (2006) suggestion that "we are not moral agents but robots are" (Sullins 2006, 27). Following what turns out to be a Kantian-influenced approach, Nadeau "claims that an action is a free action if and only if it is based on reasons fully thought out by the agent" (Sullins 2006, 27). Because human beings are not fully constituted rational beings but often make decisions based on emotional attachments and prejudices, only a logically directed machine would be capable of being a moral agent.

The fourth viewpoint, and the one that Sullins pursues, is derived from the work of Luciano Floridi and J. W. Sanders (2004), who introduce the concept of "mindless morality." "The way around the many apparent paradoxes in moral theory," Sullins (2006, 27) explains, "is to adopt a 'mindless morality' that evades issues like free will and intentionality since these are all unresolved issues in the philosophy of mind." Toward this end, Sullins proposes to redefine moral agency as involving just three criteria:

1. Autonomy, in the "engineering sense" "that the machine is not under the direct control of any other agent or user."

2. Intentionality, understood in the "weak sense" that Dennett (1998, 7) develops in his essay "Intentional systems," whereby it is not necessary to know whether some entity "*really* has beliefs and desires" but that "one can explain and predict their behaviour by *ascribing* beliefs and desires to them."

3. Responsibility, which also skirts the "other minds problem" by being satisfied with mere appearances and purposefully putting to the side the big but ultimately unresolvable metaphysical quandaries (Sullins 2006, 28).

This revised and entirely pragmatic characterization of moral agency, Sullins (ibid., 29) concludes, would apply not only to real-world embodied mechanisms, like robotic caregivers, but also software bots, corporations, animals, and the environment.

Although Sullins references and bases his own efforts on the work of Floridi and Sanders, the latter provide for an even more finely tuned reformulation of moral agency. According to Floridi and Sanders (2004, 350), the main problem for moral philosophy is that the field "remains unduly constrained by its anthropocentric conception of agenthood." This concept, they argue, does not scale to recent innovations like "distributed morality" where there is "collective responsibility resulting from the 'invisible hand' of systemic interactions among several agents at the local level," and "artificial agents (AAs) that are sufficiently informed, 'smart,' autonomous and able to perform morally relevant actions independently of the humans that created them" (ibid., 351). The problem, however, is not that these new forms of agency are not able to be considered agents, it is that the yardstick that has been employed to evaluate agency is already skewed by human prejudice. For this reason, Floridi and Sanders (ibid.) suggest that these problems and the debates they engender can be "eliminated by fully revising the concept of 'moral agent.'"

The revision proceeds by way of what Floridi and Sanders (2004, 354, 349) call "the method of abstraction,"[7] which formulates different levels of qualifying criteria for the way "one chooses to describe, analyse, and discuss a system and its context." As Floridi and Sanders point out, when the level of abstraction (LoA) that is operationalized for and within a particular field of debate is not explicitly articulated, there is equivocation and "things get messy" (ibid., 353). In order to resolve this, they advance an explicit LoA for moral agency that includes the three following criteria: interactivity, autonomy, and adaptability.

a) Interactivity means that the agent and its environment (can) act upon each other. Typical examples include input or output value, or simultaneous engagement of an action by both agent and patient—for example gravitation forces between bodies.

b) Autonomy means that the agent is able to change state without direct response to interaction: it can perform internal transitions to change its state. So an agent must have a least two states. This property imbues an agent with a certain degree of complexity and independence from its environment.

c) Adaptability means that the agent's interactions (can) change the transition rules by which it changes state. This property ensures that an agent might be viewed at the given LoA, as learning its own mode of operation in a way which depends critically on its experience. Note that if an agent's transition rules are stored as part of its internal state, discernible at this LoA, then adaptability follows from the other two conditions. (Ibid., 357–358)

At this LoA, human beings including human children, webbots and software agents like spam filters, organizations and corporations, and many different kinds of animals—more than would be allowed by either Singer or Regan—would all qualify for agency. But this is not yet "moral agency." In order to specify this additional qualification, Floridi and Sanders introduce the following modification: "An action is said to be morally qualifiable if and only if it can cause moral good or evil. An agent is said to be a moral agent if and only if it is capable of morally qualifiable action" (ibid., 364). What is important about this stipulation is that it is entirely phenomenological. That is, it is "based only on what is specified to be observable and not on some psychological speculation" (ibid., 365). An agent is a moral agent if its observed actions, irrespective of motivation or intentionality, have real moral consequences. Understood in this fashion, Floridi and Sanders advance a characterization of moral agency that does not necessarily require intelligence, intentionality, or consciousness. It is, as they call it, a kind of "mindless morality," which is something similar to Rodney Brooks's (2002, 121) "dumb, simple robots" that exhibit what appears to be intelligent behavior without necessarily possessing what is typically considered cognition or reason. "On this view," Wallach and Allen (2009, 203) write, "artificial agents that satisfy the criteria for inter-activity, autonomy, and adaptability are legitimate, fully accountable sources of moral (or immoral) actions, even if they do not exhibit free will, mental states, or responsibility."

Although providing greater precision in the characterization of the concept of moral agency and, in the process, opening up the community of moral subjects to a wider number of possible participants, Floridi and Sanders's proposal has at least three critical problems. The first has to do with equivocations that both underlie and threaten to undermine their own efforts at terminological rigor. This is the root of Johnson and Miller's (2008) critique. In particular, Johnson and Miller are concerned that the "method of abstraction," although useful insofar as it "allows us to focus on some details while ignoring other details" (Johnson and Miller 2008, 132), unfortunately permits and even facilitates significant terminological slippage. They worry, for instance, that what we call "autonomous" at one level of abstraction is not necessarily the same as "autonomous" as it is operationalized at another, and that the use of the same word in two entirely different contexts could lead to passing from the one to the other without recognizing the transference. "Our point," Johnson and Miller conclude, "is that it is misleading and perhaps deceptive to uncriti-cally transfer concepts developed at one level of abstraction to another level of abstraction. Obviously, there are levels of abstraction in which computer behavior appears autonomous, but the appropriate use of the term 'autonomous' at one level of abstraction does not mean that com-puter systems are therefore 'autonomous' in some broad and general sense" (ibid.).

In advancing this argument, however, Johnson and Miller appear to have missed the point. Namely, if the LoA is not specified, which has all too often been the case with moral agency, such slippage does and will occur. It is only by way of specifying the LoA—that is, being explicit about context and the way a particular term comes to be operationalized—that one can both avoid and protect against this very problem. In other words, what Johnson and Miller target in their critique of the method of abstrac-tion is exactly what Floridi and Sanders take as its defining purpose and raison d'être. There are, however, additional and more significant problems. In particular, the choice of a particular LoA for moral agency is clearly an important and crucial decision, but there is, it seems, some disagreement and equivocation concerning the list of qualifying criteria. Floridi and Sanders set the level at interaction, autonomy, and adaptability. But Sullins (2006), who follows their approach and also utilizes the method of abstraction, sets the LoA differently, arguing that it should include autonomy, intentionality, and understanding responsibility. From the perspective of Floridi and Sanders, Sullins sets the bar too high; from the perspective of Sullins, Floridi and Sanders set the bar too low. Although operating without an explicit LoA can be, in the words of Floridi and Sanders, "messy," operating with one is no less messy insofar as the specific LoA appears to be contentious, uncertain, and debatable. Instead of stabilizing things so as "to allow a proper definition" (Floridi and Sanders 2004, 352), the method of abstraction perpetuates the dispute and is, in the final analysis, just as "messy."

Second, and following from this, the method of abstraction, although having the appearance of an objective science modeled on "the discipline of mathematics"[8] (ibid., 352), has a political–ethical dimension that is neither recognized nor examined by Floridi and Sanders. Whoever gets to introduce and define the LoA occupies a very powerful and influential position, one that, in effect, gets to decide where to draw the line dividing "us from them." In this way, then, the method of abstraction does not really change or affect the standard operating presumptions of moral philosophy or the rules of its game. It also empowers someone or something to decide who or what is included in the community of moral subjects and who or what is to be excluded and left on the outside. And this decision, as Johnson and Miller (2008, 132) correctly point out, is something that human beings have already bestowed on themselves. If *we* decide to deploy one LoA, we exclude machines and animals, whereas another LoA allows for some animals but not machines, and still another allows some machines but not animals, and so on. It matters who gets to make these decisions, how they come to be instituted, and on what grounds, as the history of moral philosophy makes abundantly clear. "Some *definienda*," as Floridi and Sanders (2004, 353) point out, "come pre-formatted by transparent LoAs. . . . Some other *definienda* require explicit acceptance of a given LoA as a precondition for their analysis." Although Floridi and Sanders recognize that agenthood is of the latter variety, they give little consideration to the political or moral dimensions of this "explicit acceptance" or the particularly influential position they have already given themselves in this debate. They are not just diagnosing a problem from the outside but are effectively shaping its very condition for possibility. And this occupation, whether it is ever explicitly recognized as such or not, is already a moral decision. That is, it proceeds from certain normative assumptions and has specific ethical consequences.

Finally, and what makes things even more convoluted, Floridi and Sanders do not consider these difficulties and therefore effectively avoid responding to and taking responsibility for them. The LoA approach, as Floridi and Sanders describe it, is not designed to define moral agency, but merely to provide operational limits that can be used to help decide whether something has achieved a certain benchmark threshold for inclusion or not. "We clarify the concept of moral agent," Floridi and Sanders write, "by providing not a definition but an effective characterization, based on three criteria at a specified LoA" (ibid., 351). There is, of course, nothing inherently wrong with this entirely pragmatic and practical approach. It does, however, apply what is arguably an "engineering solution" to a fundamental philosophical problem. Instead of advancing and defending a decision concerning a definition of moral agency, Floridi and Sanders only advance an "effective characterization" that can work. In their estimation, therefore, what they do is situated beyond good and evil; it is simply an expedient and instrumental way to address and dispense with moral agency. So it appears that Heidegger got it right in his 1966 *Der Spiegel* interview when he suggested that the science and engineering practices of cybernetics have in recent years taken the place of what had been called philosophy (Heidegger 2010, 59). This kind of functionalist approach, although entirely useful as demonstrated by Floridi and Sanders, has its own costs as well as benefits (which is, it should be noted, an entirely functionalist way to address the matter).

1.5.2 Functional Morality

Attempts to resolve the question of moral agency run into considerable metaphysical, epistemological, and moral difficulties. One way to work around these problems is to avoid the big philosophical questions altogether. This is precisely the strategy utilized by engineers who advocate a functionalist approach or "applications route" (Schank 1990, 7). This alternative strategy, what Wendell Wallach (2008, 466) calls a "functional morality," recognizes that machine agency might not be decidable but that this undecidability is no excuse for not considering the real-world consequences of increasingly autonomous machine decision making. As Susan and Michael Anderson (2007a, 16) explain, "there are ethical ramifications to what machines currently do and are projected to do in the future. To neglect this aspect of machine behavior could have serious repercussions." In other words, while we busy ourselves with philosophical speculation concerning the moral status of the machine, machines are already making decisions that might have devastating effects for us and our world. So rather than quibbling about obscure metaphysical details or epistemological limitations that might exceed our ability to judge, we should work with and address the things to which we do have access and can control.

This alternative transaction, which bears an uncanny resemblance to Kant's critical endeavors,[9] attempts to address the practical matter of moral responsibility without first needing to entertain or resolve the big metaphysical, epistemological, ontological, or metaethical questions. This does not mean, it is important to note, that one either accepts or denies the question of machine agency, personhood, or consciousness. Instead it merely suggests that we take a Kantian critical stance, recognizing that this question in and of itself may exceed our limited capacities. So instead of trying to solve the seemingly irreducible problem of "other minds," the functionalist simply decides not to decide. In this way, functionalism remains agnostic about the state of machine consciousness, for example, and endeavors to pursue the subject in a much more practical and utilitarian manner.

There have been various attempts at instituting this kind of functionalist approach. The first and perhaps best-known version of it is Isaac Asimov's "laws of robotics." These three laws[10] are behavioral rules that are designed to restrict programmatically robotic action.

1. A robot may not injure a human being or, through inaction, allow a human being to come to harm.

2. A robot must obey any orders given to it by human beings, except where such orders would conflict with the First Law.

3. A robot must protect its own existence as long as such protection does not conflict with the First or Second Laws. (Asimov 2008, 37)

These laws are entirely functional. That is, they do not necessarily require (nor do they preclude) a decision concerning machinic consciousness and personhood. They are simply program instructions that are designed and intended to regulate actual robotic actions. As Wendell Wallach points out in an article initially published in *AI & Society*, this is just good engineering practice: "Engineers have always been concerned with designing tools that are safe and reliable. Sensitivity to the moral implications of two or more courses of action in limited contexts can be understood as an extension of the engineer's concern with designing appropriate control mechanisms for safety into computers and robots" (Wallach 2008, 465).

Although possessing considerable promise for a functionalist and very pragmatic approach to the problem, Asimov's laws have been criticized as insufficient and impractical. First, Asimov himself employed the laws not to solve problems with machine action and behavior but to generate interesting science fiction stories. Consequently, Asimov did not intend the rules to be a complete and definitive set of instructions for robots but used the laws as a literary device for generating dramatic tension, fictional scenarios, and character conflicts. As Lee McCauley (2007, 160) succinctly explains, "Asimov's Three Laws of Robotics are literary devices and not engineering principles." Second, theorists and practitioners working the fields of robotics and computer ethics have found Asimov's laws to be underpowered for everyday practical employments. Susan Leigh Anderson, for example, directly grapples with this issue in the essay "Asimov's 'Three Laws of Robotics' and Machine Metaethics," demonstrating not only that "Asimov rejected his own three laws as a proper basis for machine ethics" (Anderson 2008, 487) but that the laws, although providing a good starting point for discussion and debate about the matter, "are an unsatisfactory basis for machine ethics" (ibid., 493). Consequently, "even though knowledge of the Three Laws of Robotics seem universal among AI researchers," McCauley (2007, 153) concludes, "there is the pervasive attitude that the Laws are not implementable in any meaningful sense."

Despite these misgivings, the functionalist approach that is modeled by Asimov's three laws is not something that is limited to fiction. It also has very real applications. Perhaps the most ambitious effort in this area has been in the field of machine ethics (ME). This relatively new idea was first introduced and publicized in a paper written by Michael Anderson, Susan Leigh Anderson, and Chris Armen and presented during the 2004

Workshop on Agent Organizations held in conjunction with the American Association for Artificial Intelligence's (AAAI) nineteenth national conference. This debut, which appropriately sought "to lay the theoretical foundation for *machine ethics*" (Anderson, Anderson, and Armen 2004, 1) was quickly followed with the formation of the Machine Ethics Consortium (MachineEthics.org), a 2005 AAAI symposium on the subject, and a dedicated issue of *IEEE Intelligent Systems* published in the summer of 2006. Unlike computer ethics, which is mainly concerned with the consequences of human behavior through the instrumentality of computer technology, "*machine ethics* is concerned," as characterized by Anderson et al., "with the consequences of behavior of machines toward human users and other machines" (ibid.). In this way, machine ethics both challenges the "human-centric" tradition that has persisted in moral philosophy and argues for a widening of the subject of ethics so as to take into account not only human action with machines but the behavior of some machines, namely those that are designed to provide advice or programmed to make autonomous decisions with little or no human supervision.

Toward this end, machine ethics takes an entirely functionalist approach. That is, it considers the effect of machine actions on human subjects irrespective of metaphysical debates concerning agency or epistemological problems concerning subjective mind states. As Susan Leigh Anderson (2008, 477) points out, the ME project is unique insofar as it, "unlike creating an autonomous ethical machine, will not require that we make a judgment about the ethical status of the machine itself, a judgment that will be particularly difficult to make." The project of machine ethics, therefore, does not necessarily deny or affirm the possibility of machine consciousness and personhood. It simply endeavors to institute a pragmatic approach that does not require that one first decide this ontological question a priori. ME therefore leaves this as an open question and proceeds to ask whether moral decision making is computable and whether machines can in fact be programmed with appropriate ethical standards for behavior.

In response to the first concern—whether ethics is computable—it should be noted that moral philosophy has often been organized according to a mechanistic or computational model. This goes not only for act utilitarianism, which is the ethical theory to which the Andersons are drawn, but also its major competitor in modern philosophy—deontologism. Both utilitarianism and Kantian deontological ethics strive for a rational mechanization of moral decision making. In fact, the mechanical aspect of moral reasoning has been celebrated precisely because it removes any and all emotional investments that could cause capricious and unjust decision making. According to Henry Sidgwick (1981, 77), for example, "the aim of Ethics is *to systematize* and free from error the apparent cognitions that most men have of the rightness or reasonableness of conduct." Consequently, Western conceptions of morality customarily consist in systematic rules of behavior that can be encoded, like an algorithm, and implemented by different moral agents in a number of circumstances and situations. They are, in short, program instructions that are designed to direct behavior and govern conduct. Take, for instance, the Ten Commandments, the cornerstone of Judeo-Christian ethics. These ten rules constitute an instruction set that not only prescribes correct operations for human beings but does so in a way that is abstracted from the particulars of circumstance, personality, and other empirical accidents. "Thou shall not kill" is a general prohibition against murder that applies to any number of situations where one human being confronts another. Like an algorithm, the statements contained within the Ten Commandments are general operations that can be applied to any particular set of data.

Similarly, Kant's moral philosophy is founded on and structured by fundamental rules or what he calls, in a comparison to the laws of natural science, "practical laws." These practical laws are "categorical imperatives." That is, they are not merely subjective maxims that apply to a particular person's will under a specific set of circumstances. Instead, they must be objectively valid for the will of every rational being in every possible circumstance. "Laws," Kant (1985, 18) writes, "must completely determine the will as will, even before I ask whether I am capable of achieving a desired effect or what should be done to realize it. They must thus be categorical; otherwise they would not be laws, for they would lack the necessity which, in order to be practical, must be completely independent of pathological conditions, i.e., conditions only contingently related to the will." For Kant, moral action is programmed by principles of pure practical reason—universal laws that are not only abstracted from every empirical condition but applicable to any and all rational agents. It may be said, therefore, that Kant, who took physics and mathematics as the model for a wholesale transformation of the procedures of philosophy, mechanized ethics in a way that was similar to Newton's mechanization of physical science.

Finally, even the pragmatic alternative to deontological ethics, utilitarianism, operates by a kind of systemic moral computation or what Jeremy Bentham called "moral arithmetic" (Dumont 1914, 2). The core utilitarian principle, "seek to act in such a way as to promote the greatest quantity and quality of happiness for the greatest number," is a general formula that subsequently requires considerable processing to crunch the numbers and decide the best possible outcome. For this reason, Anderson and Anderson (2007b, 5) have suggested that "computers might be better at following an ethical theory than most humans," because humans "tend to be inconsistent in their reasoning" and "have difficulty juggling the complexities of ethical decision-making" owing to the sheer volume of data that need to be taken into account and processed.

The question "Is ethics computable?" comprises, as Anderson and Anderson (2007b, 5) point out, "the central question of the Machine Ethics project." In order to respond to it, Anderson and Anderson (2007a, 22), following the hacker adage that "programming is argument," have designed several working prototypes "to demonstrate the possibility of creating a machine that is an explicit ethical agent." Their first projects consisted in two computerized ethical advisors, *Jeremy*, which was based on an implementation of Bentham's act utilitarianism, and *W.D.*, which was designed to apply W. D. Ross's (2002) deontological ethics of prima facie duties (Anderson, Anderson, and Armen 2004). These initial projects have been followed by MedEthEx, an expert system "that uses machine learning to resolve a biomedical ethical dilemma" (Anderson, Anderson, and Armen 2006) and EthEl, "a system in the domain of elder care that determines when a patient should be reminded to take mediation and when a refusal to do so is serious enough to contact an overseer" (Anderson and Anderson 2007a, 24). Although both systems are designed around an implementation of Beauchamp and Childress's (1979) four principles of biomedical ethics, EthEl is designed to be more autonomous. "Whereas MedEthEx," Anderson and Anderson (2007a, 24) write, "gives the ethically correct answer (that is, that which is consistent with its training) to a human user who will act on it or not, EthEl herself acts on what she determines to be the ethically correct action."

Whether this approach will eventually produce an "ethical intelligent agent" is something that has yet to be seen. For now, what the Andersons and their collaborators have demonstrated, through "proof of concept applications in constrained domains," is that it is possible to incorporate explicit ethical components in a machine (ibid., 25). This is, Anderson and Anderson (2007a) conclude, not only an important engineering accomplishment but something that could potentially contribute to advancements in moral theory. This is because moral philosophy, in their estimation at least, has been a rather imprecise, impractical, and error-prone undertaking. By making ethics computable and proving this by way of working demonstrations, ME will not only "discover problems with current theories" but might even lead to the development of better theories. "It is important," Anderson and Anderson write, perhaps with reference to their own collaborative endeavors, "to find a clear, objective basis for ethics—making ethics in principle computable—if only to rein in unethical human behavior; and AI researchers, working with ethicists, have a better chance of achieving breakthroughs in ethical theory than theoretical ethicists working alone" (ibid., 17).

A similar functionalist approach is instituted by Wendell Wallach and Colin Allen (2009, 58), who admit that deciding machine consciousness will most likely remain an open question. Although Wallach and Allen recognize the importance of the more profound and perhaps ultimately irresolvable philosophical questions, this fact does not stop them from advocating for the design of systems that have some functional moral capacity. Toward this end, Colin Allen, Gary Varner and Jason Zinser (2000, 251) introduced the term "artificial moral agent" (AMA) for these "future systems and software agents sensitive to moral considerations in the execution of their tasks, goals, and duties" (Wallach 2008, 466). Developing functional AMAs will, according to Wallach's experience, entail productive collaboration and dialogue between philosophers, who "are knowledgeable about the values and limits inherent in the various ethical orientations," and engineers, who "understand what can be done with existing technologies and those technologies we will witness in the near future" (ibid.). From this perspective, Wallach and Allen, first in a conference paper called "Android Ethics" (2005) and then in the book *Moral Machines* (2009) propose and pursue a "cost-benefit analysis" of three different approaches to designing functional AMAs—top-down, bottom-up, and hybrid.

The top-down approach, Wallach and Allen (2005, 150) explain, "combines two slightly different senses of this term, as it occurs in engineering and as it occurs in ethics." In its merged form, "a top-down approach to the design of AMAs is any approach that takes the antecedently specified ethical theory and analyzes its computational requirements to guide the design of algorithms and subsystems capable of implementing that theory" (ibid., 151). This is the approach to AMA design that is exemplified by Asimov's three laws and has been implemented in the work of Selmer Bringsjord's at Rensselar Polytechnic Institute's AI and Reasoning Laboratory. This "rigorous, logic-based approach to software engineering requires AMA designers to formulate, up front, consistent ethical code for any situation where they wish to deploy an AMA" (Wallach and Allen 2009, 126). According to Wallach and Allen's analysis, however, this inductive approach to moral reasoning and decision making is limited and only really works in situations that are carefully controlled and highly restricted. Consequently, "the limitations of top-down approaches nevertheless add up, on our view, to the conclusion that it will not be feasible to furnish an AMA with an unambiguous set of top-down rules to follow" (ibid., 97).

The bottom-up approach, as its name indicates, proceeds in the opposite direction. Again, "bottom-up" is formulated a bit differently in the fields of engineering and moral philosophy.

In bottom-up engineering tasks can be specified *a*theoretically using some sort of performance measure. Various trial and error techniques are available to engineers for progressively tuning the performance of systems so that they approach or surpass performance criteria. High levels of performance on many tasks can be achieved, even though the engineer lacks a theory of the best way to decompose the tasks into subtasks. . . . In its ethical sense, a bottom-up approach to ethics is one that treats normative values as being implicit in the activity of agents rather than explicitly articulated in terms of a general theory. (Wallach and Allen 2005, 151)

The bottom-up approach, therefore, derives moral action from a kind of trial-and-error process where there is no resolution of or need for any decision concerning a general or generalizable theory. Theory might be able to be derived from such trials, but that is neither necessary nor required. This is, then, a kind of deductive approach, and it is exemplified by Peter Danielson's (1992) "virtuous robots for virtual games" and the Norms Evolving in Response to Dilemmas (NERD) project. In these situations, morality is not something prescribed by a set of preprogrammed logical rules to be applied but "emerges out of interactions among multiple agents who must balance their own needs against the competing demands of others" (Wallach and Allen 2009, 133). This approach, Wallach and Allen explain, has the distinct advantage that it "focuses attention on the social nature of ethics" (ibid.). At the same time, however, it is unclear, in their estimation at least, how such demonstrations would scale to larger real-world applications.

The top-down and bottom-up approaches that Wallach and Allen investigate in their consideration of AMA design are not unique. In fact, they parallel and are derived from the two main strategies undertaken in and constitutive of the field of AI (Brooks 1999, 134). "In top-down AI," Jack Copland (2000, 2) writes, "cognition is treated as a high-level phenomenon that is independent of the low-level details of the implementing mechanism—a brain in the case of a human being, and one or another design of electronic digital computer in the artificial case. Researchers in bottom-up AI, or *connectionism*, take an opposite approach and simulate networks of artificial neurons that are similar to the neurons in the human brain. They then investigate what aspects of cognition can be recreated in these artificial networks." For Wallach and Allen (2009, 117), as for many researchers in the field of AI, "the top-down/bottom-up dichotomy is somewhat simplistic." Consequently, they advocate hybrid approaches that combine the best aspects and opportunities of both. "Top-down analysis and bottom-up techniques for developing or evolving skills and mental faculties will undoubtedly both be required to engineer AMAs" (Wallach and Allen 2005, 154).

Although fully operationally hybrid AMAs are not yet available, a number of research projects show considerable promise. Wallach and Allen, for instance, credit Anderson and Anderson's MedEthEx, which employs both predefined prima facie duties and learning algorithms, as a useful, albeit incomplete, implementation of this third way. "The approach taken by the Andersons," Wallach and Allen (2009, 128) write, "is almost completely top-down—the basic duties are predefined, and the classification of cases is based on those medical ethicists generally agree on. Although MedEthEx learns from cases in what might seem in a sense to be a 'bottom-up' approach, these cases are fed into the learning algorithm as high-level descriptions using top-down concepts of the various duties that may be satisfied or violated. The theory is, as it were, spoon fed to the system rather than it having to learn the meaning of 'right' and 'wrong' for itself." Taking things one step further is Stan Franklin's learning intelligent distribution agent or LIDA. Although this conceptual and computational model of cognition was not specifically designed for AMA development, Wallach and Allen (2009, 172) find its systems architecture, which can "accommodate top-down analysis and bottom up propensities," to hold considerable promise for future AMA design.

Despite promising results, the functionalist approach has at least three critical difficulties. The first has to do with testing. Once "moral functionalism," as Danielson (1992, 196) calls it, is implemented, whether by way of utilizing top-down, bottom-up, or some hybrid of the two, researchers will need some method to test whether and to what extent the system actually works. That is, we will need some metric by which to evaluate whether or not a particular device is capable of making the appropriate moral decisions in a particular situation. Toward this end, Allen, Varner, and Zinser (2000) introduce a modified version of the Turing test which they call the moral Turing test (MTT).

In the standard version of the Turing Test, an "interrogator" is charged with distinguishing a machine from a human based on interacting with both via printed language alone. A machine passes the Turing Test if, when paired with a human being, the "interrogator" cannot identify the human at a level above chance. . . . A Moral Turing Test (MTT) might similarly be proposed to bypass disagreements about ethical standards by restricting the standard Turing Test to conversations about morality. If human "interrogators" cannot identify the machine at above chance accuracy, then the machine is, on this criterion, a moral agent. (Allen, Varner, and Zinser 2000, 254)

The moral Turing test, therefore, does not seek to demonstrate machine intelligence or consciousness or resolve the question of moral personhood. It merely examines whether an AMA can respond to questions about moral problems and issues in a way that is substantively indistinguishable from a human moral agent.

This method of testing has the advantage that it remains effectively agnostic about the deep metaphysical questions of personhood, person-making properties, and the psychological dimensions typically associated with agency. It is only interested in demonstrating whether an entity can pass as a human-level moral agent in conversation about ethical matters or evaluations of particular ethical dilemmas. At the same time, however, the test has been criticized for the way it places undue emphasis on the discursive abilities "to *articulate* moral judgments" (ibid.). As Stahl (2004, 79) points out, "in order to completely participate in a dialogue that would allow the observer or 'interrogator' to determine whether she is dealing with a moral agent, the computer would need to understand the situation in question." And that means that the computer would not simply be manipulating linguistic symbols and linguistic tokens concerning moral subjects, but "it would have to understand a language," which is, according to Stahl's understanding, "something that computers are not capable of" (ibid., 80). Although bypassing metaphysical speculation, the MTT cannot, it seems, escape the requirements and complications associate with λόγος. Consequently, this apparently practical test of moral functionalism ultimately reinstates and redeploys the theoretical problems the functionalist approach was to have circumvented in the first place.

Second, functionalism shifts attention from the cause of a moral action to its effects. By remaining effectively agnostic about personhood or consciousness, moral questions are transferred from a consideration of the intentionality of the agent to the effect an action has on the recipient, who is generally assumed to be human. This presumption of human patiency is immediately evident in Asimov's three laws of robotics, which explicitly stipulate that a robot may not, under any circumstance, harm a human being. This anthropocentric focus is also a guiding principle in AMA development, where the objective is to design appropriate safeguards into increasingly complex systems. "A concern for safety and societal benefits," Wallach and Allen (2009, 4) write, "has always been at the forefront of engineering. But today's systems are approaching a level of complexity that, we argue, requires the systems themselves to make moral decisions— to be programmed with 'ethical subroutines,' to borrow a phrase from *Star Trek*. This will expand the circle of moral agents beyond humans to artificially intelligent systems, which we call artificial moral agents." And the project of machine ethics proceeds from and is interested in the same. "Clearly," Anderson and company write (2004, 4), "relying on machine intelligence to effect change in the world without some restraint can be dangerous. Until fairly recently, the ethical impact of a machine's actions has either been negligible, as in the case of a calculator, or, when considerable, has only been taken under the supervision of a human operator, as in the case of automobile assembly via robotic mechanisms. As we increasingly rely upon machine intelligence with reduced human supervision, we will need to be able to count on a certain level of ethical behavior from them." The functionalist approaches, therefore, derive from and are motivated by an interest to protect human beings from potentially hazardous machine decision making and action. Deploying various forms of machine intelligence and autonomous decision making in the real world without some kind of ethical restraint or moral assurances is both risky and potentially dangerous for human beings. Consequently, functionalist approaches like that introduced by the Andersons' machine ethics, Asimov's three laws of robotics, or Wallach and Allen's *Moral Machines*, are motivated by a desire to manage the potential hazards of machine decision making and action for the sake of ensuring the humane treatment of human beings.

This has at least two important consequences. On the one hand, it is thoroughly and unapologetically anthropocentric. Although effectively opening up the community of moral agents to other, previously excluded subjects, the functionalist approach only does so in an effort to protect human interests and investments. This means that the project of machine ethics or machine morality do not differ significantly from computer ethics and its predominantly anthropocentric orientation. If computer ethics, as Anderson, Anderson, and Armen (2004) characterize it, is about the responsible and irresponsible use of computerized tools by human agents, then the functionalist approaches are little more than the responsible programming of machines by human beings for the sake of protecting other human beings. In some cases, like Wallach and Allen's machine morality, this anthropocentrism is not necessarily a problem or considered to be a significant concern. In other cases, however, it does pose significant difficulties. Machine ethics, for example, had been introduced and promoted as a distinct challenge to the anthropocentric tradition in general and as an alternative to the structural limitations of computer ethics in particular (ibid., 1). Consequently, what machine ethics explicitly purports to do might be in conflict with what it actually does and accomplishes. In other words, the critical challenge machine ethics advances in response to the anthropocentric tradition in computer ethics is itself something that is mobilized by and that ultimately seeks to protect the same fundamental anthropocentric values and assumptions.

On the other hand, functionalism institutes, as the conceptual flip side and consequence of this anthropocentric privilege, what is arguably a slave ethic. "I follow," Kari Gwen Coleman (2001, 249) writes, "the traditional assumption in computer ethics that computers are merely tools, and intentionally and explicitly assume that the end of computational agents is to serve humans in the pursuit and achievement of their (i.e. human) ends.

In contrast to James Gips' call for an ethic of equals, then, the virtue theory that I suggest here is very consciously a slave ethic." For Coleman, computers and other forms of computational agents should, in the words of Bryson (2010), "be slaves." In fact, Bryson argues that treating robots and other autonomous machines in any other way would be both inappropriate and unethical. "My thesis," Bryson writes, "is that robots should be built, marketed and considered legally as slaves, not companion peers" (ibid., 63).

Others, however, are not so confident about the prospects and consequences of this "Slavery 2.0." And this concern is clearly one of the standard plot devices in robot science fiction from *R.U.R.* and *Metropolis* to *Bladerunner* and *Battlestar Galactica*. But it has also been expressed by contemporary researchers and engineers. Rodney Brooks, for example, recognizes that there are machines that are and will continue to be used and deployed by human users as instruments, tools, and even servants. But he also recognizes that this approach will not cover all machines.

Fortunately we are not doomed to create a race of slaves that is unethical to have as slaves. Our refrigerators work twenty-four hours a day seven days a week, and we do not feel the slightest moral concern for them. We will make many robots that are equally unemotional, unconscious, and unempathetic. We will use them as slaves just as we use our dishwashers, vacuum cleaners, and automobiles today. But those that we make more intelligent, that we give emotions to, and that we empathize with, will be a problem. We had better be careful just what we build, because we might end up liking them, and then we will be morally responsible for their well-being. Sort of like children. (Brooks 2002, 195)

According to this analysis, a slave ethic will work, and will do so without any significant moral difficulties or ethical friction, as long as *we* decide to produce dumb instruments that serve human users as mere prostheses. But as soon as the machines show signs, however minimal defined or rudimentary, that *we take* to be intelligent, conscious, or intentional, then everything changes. At that point, a slave ethic will no longer be functional or justifiable; it will become morally suspect.

Finally, even those seemingly unintelligent and emotionless machines that can legitimately be utilized as "slaves" pose a significant ethical problem. This is because machines that are designed to follow rules and operate within the boundaries of some kind of programmed restraint might turn out to be something other than what is typically recognized as a moral agent. Terry Winograd (1990, 182–183), for example, warns against something he calls "the bureaucracy of mind," "where rules can be followed without interpretive judgments." Providing robots, computers, and other autonomous machines with functional morality produces little more than artificial bureaucrats—decision-making mechanisms that can follow rules and protocols but have no sense of what they do or understanding of how their decisions might affects others. "When a person," Winograd argues, "views his or her job as the correct application of a set of rules (whether human-invoked or computer-based), there is a loss of personal responsibility or commitment. The 'I just follow the rules' of the bureaucratic clerk has its direct analog in 'That's what the knowledge base says.' The individual is not committed to appropriate results, but to faithful application of procedures" (ibid., 183).

Mark Coeckelbergh (2010, 236) paints an even more disturbing picture. For him, the problem is not the advent of "artificial bureaucrats" but "psychopathic robots." The term "psychopathy" has traditionally been used to name a kind of personality disorder characterized by an abnormal lack of empathy that is masked by an ability to appear normal in most social situations. Functional morality, Coeckelbergh argues, intentionally designs and produces what are arguably "artificial psychopaths"—robots that have no capacity for empathy but which follow rules and in doing so can appear to behave in morally appropriate ways. These psychopathic machines would, Coeckelbergh argues, "follow rules but act without fear, compassion, care, and love. This lack of emotion would render them non-moral agents—i.e. agents that follow rules without being moved by moral concerns—and they would even lack the capacity to discern what is of value. They would be morally blind" (ibid.).

Consequently, functionalism, although providing what appears to be a practical and workable solution to the problems of moral agency, might produce something other than artificial moral agents. In advancing this critique, however, Winograd and Coeckelbergh appear to violate one of the principal stipulations of the functionalist approach. In particular, their critical retort presumes to know something about the inner state of the machine, namely, that it lacks empathy or understanding for what it does. This is precisely the kind of speculative knowledge about other minds that the functionalist approach endeavors to remain agnostic about: one cannot ever know whether another entity does or does not possess a particular inner disposition. Pointing this out, however, does not improve or resolve things. In fact, it only makes matters worse, insofar as we are left with considerable uncertainty whether functionally designed systems are in fact effective moral agents, artificial bureaucrats coldly following their programming, or potentially dangerous psychopaths that only appear to be normal.

1.6 Summary

The machine question began by asking about moral agency, specifically whether AIs, robots, and other autonomous systems could or should be considered a legitimate moral agent. The decision to begin with this subject was not accidental, provisional, or capricious. It was dictated and prescribed by the history of moral philosophy, which has traditionally privileged agency and the figure of the moral agent in both theory and practice. As Floridi (1999) explains, moral philosophy, from the time of the ancient Greeks through the modern era and beyond, has been almost exclusively an agent-oriented undertaking. "Virtue ethics, and Greek philosophy more generally," Floridi (1999, 41) argues, "concentrates its attention on the moral nature and development of the individual agent who performs the action. It can therefore be properly described as an agent-oriented, 'subjective ethics.'" Modern developments in moral philosophy, although shifting the focus somewhat, retain this particular orientation. "Developed in a world profoundly different from the small, non-Christian Athens, Utilitarianism, or more generally Consequentialism, Contractualism and Deontologism are the three most well-known theories that concentrate on the moral nature and value of the actions performed by the agent" (ibid.). Although shifting focus from the "moral nature and development of the individual agent" to the "moral nature and value" of his or her actions, Western philosophy has been, with few exceptions (which we will get to shortly), organized and developed as an agent-oriented endeavor.

As we have seen, when considered from the perspective of the agent, moral philosophy inevitably and unavoidably makes exclusive decisions about *who* is to be included in the community of moral agents and *what* can be excluded from consideration. The choice of words is not accidental; it too is necessary and deliberate. As Derrida (2005, 80) points out, everything turns on and is decided by the difference that separates the "who" from the "what." Agency has been customarily restricted to those entities who call themselves and each other "man"—those beings who already give themselves the right to be considered someone who counts as opposed to something that does not. But who counts—who, in effect, gets to be situated under the term "who"—has never been entirely settled, and the historical development of moral philosophy can be interpreted as a progressive unfolding, where what had been excluded (women, slaves, people of color, etc.) have slowly and not without considerable struggle and resistance been granted access to the gated community of moral agents and have thereby come to be someone who counts.

Despite this progress, which is, depending on how one looks at it, either remarkable or insufferably protracted, machines have not typically been included or even considered as possible candidates for inclusion. They have been and continue to be understood as mere artifacts that are designed, produced, and employed by human agents for human-specified ends. They are, then, as it is so often said by both technophiles and technophobes, nothing more than a means to an end. This instrumentalist understanding of technology has achieved a remarkable level of acceptance and standardization, as is evidenced by the fact that it has remained in place and largely unchallenged from ancient to postmodern times—from at least Plato's *Phaedrus* to Lyotard's *The Postmodern Condition*. And this fundamental decision concerning the moral position and status of the machine—or, better put, the lack thereof—achieves a particularly interesting form when applied to autonomous systems and robots, where it is now argued, by Bryson (2010) and others, that "robots should be slaves." To put it another way, the standard philosophical decision concerning who counts as a moral agent and what can and should be excluded has the result of eventually producing a new class of slaves and rationalizing this institution as morally justified. It turns out, then, that the instrumental theory is a particularly good instrument for instituting and ensuring human exceptionalism and authority.

Despite this, beginning with (at least) Heidegger's critical intervention and continuing through both the animal rights movement and recent advancements in AI and robotics, there has been considerable pressure to reconsider the metaphysical infrastructure and moral consequences of this instrumentalist and anthropocentric legacy. Extending consideration to these other previously excluded subjects, however, requires a significant reworking of the concept of moral "personhood," one that is not dependent on genetic makeup, species identification, or some other spurious criteria. As promising as this development is, "the category of the person," to reuse terminology borrowed from Mauss's essay (Carrithers, Collins, and Lukes 1985), is by no means without difficulty. In particular, as we have seen, there is little or no agreement concerning what makes someone or something a person. Consequently, as Dennett (1998, 267) has pointed out, "person" not only lacks a "clearly formulatable necessary and sufficient conditions for ascription" but, in the final analysis, is perhaps "incoherent and obsolete."

In an effort to address if not resolve this problem, we followed the one "person making" quality that appears on most, if not all, the lists, whether they are composed of just a couple simple elements (Singer 1999, 87) or involve numerous "interactive capacities" (Smith 2010, 74), and which already has considerable traction with theorists and practitioners—consciousness. In fact, moral personhood, from Locke (1996, 170) to Himma (2009, 19), has often been determined to be dependent on consciousness as its necessary precondition. But this too ran into ontological and epistemological problems. On the one hand, we do not, it seems, have any idea what "consciousness" is. In a way that is similar to what Augustine (1963, xi–14) writes of "time," consciousness appears to be one of those concepts that we know what it is as long as no one asks us to explain what it is. Dennett (1998, 149–150), in fact, goes so far as to admit that consciousness is "the last bastion of occult properties." On the other hand, even if we were able to define consciousness or come to some tentative agreement concerning its characteristics, we lack any credible and certain way to determine its actual presence in others. Because consciousness is a property attributed to "other minds," its presence or lack thereof requires access to something that is and remains inaccessible. And the supposed solutions for these problems, from reworkings and modifications of the Turing test to functionalist approaches that endeavor to work around the problem of other minds altogether, only make things worse.

Responding to the question of machine moral agency, therefore, has turned out to be anything but simple or definitive. This is not, it is important to point out, because machines are somehow unable to be moral agents; it is rather a product of the fact that the term "moral agent," for all its importance and expediency, remains an ambiguous, indeterminate, and rather noisy concept. What the examination of the question of machine moral agency demonstrates, therefore, is something that was not anticipated or necessarily sought. What has been discovered in the process of pursuing this line of inquiry is not an answer to the question of whether machines are or are not moral agents. In fact, that question remains unanswered. What has been discovered is that the concept of moral agency is already so thoroughly confused and messy that it is now unclear whether we—whoever this "we" includes—are in fact moral agents. What the machine question demonstrates, therefore, is that the question concerning agency, the question that had been assumed to be the "correct" place to begin, turns out to be inconclusive. Although this could be called a failure, it is a particularly instructive failing, like any "failed experiment" in the empirical sciences. What is learned from this failure—assuming we continue to use this obviously "negative" word—is that moral agency is not necessarily something that is to be discovered in others prior to and in advance of their moral consideration. Instead, it is something that comes to be conferred and assigned to others in the process of our interactions and relationships with them. But then the issue is no longer one of agency; it is a matter of *patiency*.

2 Moral Patiency

2.1 Introduction

A patient-oriented ethics looks at things from the other side—in more ways than one. The question of moral patiency is, to put it rather schematically, whether and to what extent robots, machines, nonhuman animals, extra-terrestrials, and so on might constitute an *other* to which or to whom one would have appropriate moral duties and responsibilities. And when it comes to this particular subject, especially as it relates to artificial entities and other forms of nonhuman life, it is perhaps Mary Shelley's *Frankenstein* that provides the template. In the disciplines of AI and robotics, but also any field that endeavors to grapple with the opportunities and challenges of technological innovation, Shelley's narrative is generally considered to have instituted an entire "genre of cautionary literature" (Hall 2007, 21), that Isaac Asimov (1983, 160) termed "the Frankenstein Complex." "The story's basic and familiar outline," Janice Hocker Rushing and Thomas S. Frentz (1989, 62) explain, "is that a technologically created being appears as a surprise to an unsuspecting and forgetful creator. . . . The maker is then threatened by the made, and the original roles of master and slave are in doubt. As Shelley acknowledges by subtitling her novel *A Modern Prometheus*, Dr. Frankenstein enters forbidden territory to steal knowledge from the gods, participates in overthrowing the old order, becomes a master of technics, and is punished for his transgression."

Although this is a widely accepted and rather popular interpretation, it is by no means the only one or even a reasonably accurate reading of the text. In fact, Shelley's novel is not so much a cautionary tale warning modern scientists and technicians of the hubris of unrestricted research and the dangerous consequences of an artificial creation run amok, but a meditation on how one responds to and takes responsibility for others—especially when faced with other kinds of otherness. At a pivotal moment in the novel, when Victor Frankenstein finally brings his creature to life, it is the brilliant scientist who recoils in horror at his own creation, runs away from the scene, and abandons the creature to fend for itself. As Langdon Winner (1977, 309) insightfully points out in his attempt to reposition the story, "this is very clearly a flight from responsibility, for the creature is still alive, still benign, left with nowhere to go, and, more important, stranded with no introduction to the world in which he must live." What Shelley's narrative illustrates, therefore, is the inability of Victor Frankenstein to respond adequately and responsibly to his creation—this other being who confronts him face to face in the laboratory. The issue addressed by the novel, then, is not solely the hubris of a human agent who dares to play god and gets burned as a consequence, but the failure of this individual to respond to and to take responsibility for this other creature. The problem, then, is not necessarily one of moral agency but of *patiency*.

2.2 Patient-Oriented Approaches

The term "moral patient" does not have the same intuitive recognition and conceptual traction as its other. This is because "the term *moral patient*," Mane Hajdin (1994, 180) writes, "is coined by analogy with the term *moral agent*. The use of the term *moral patient* does not have such a long and respectable *history* as that of the term *moral agent*, but several philosophers have already used it." Surveying the history of moral philosophy, Hajdin argues that "moral patient" is not an originary term but is formulated as the dialectical flip side and counterpart of agency. For this reason, moral patiency, although recently garnering considerable attention in both analytic and continental ethics,[1] has neither a long nor a respectable history. It is a derived concept that is dependent on another, more originary term. It is an aftereffect and by-product of the agency that has been attributed to the term "moral agent." A similar explanation is provided by Tom Regan, who also understands moral patiency as something derived from and dependent on agency. "Moral agents," Regan (1983, 152) writes, "not only can do what is right or wrong, they may also be on the receiving end, so to speak, of the right and wrong acts of other moral agents. There is, then, a sort of reciprocity that holds between moral agents. . . . An individual who is not a moral agent stands outside the scope of direct moral concern on these views, and no moral agent can have any direct duty to such individuals. Any duties involving individuals who are not moral agents are indirect duties to those who are." On this view, moral patiency is just the other side and conceptual opposite of moral agency. This "standard position," as Floridi and Sanders (2004, 350) call it, "maintains that all entities that qualify as moral agents also qualify as moral patients and vice versa."

According to this "standard position," then, anything that achieves the status of moral agent must in turn be extended consideration as a moral patient. The employment of this particular logical structure in research on AI, robotics, and ethics has resulted in two very different lines of argument and opposing outcomes. It has, on the one hand, been used to justify the exclusion of the machine from any consideration of moral patiency altogether. Joanna Bryson (2010), for example, makes a strong case against ascribing moral agency to machines and from this "fact" immediately and without further consideration also denies such artifacts any access to patiency. "We should never," Bryson writes in that imperative form which is recognizably moral in tone, "be talking about machines taking ethical decisions, but rather machines operated correctly within the limits we set for them" (Bryson 2010, 67). Likewise, we should, she continues, also resist any and all efforts to ascribe moral patiency to what are, in the final analysis, mere artifacts and extensions of our own faculties. In other words, robots should be treated as tools or instruments, and as such they should be completely at our disposal, like any other object. "A robot can," Bryson argues, "be abused just as a car, piano, or couch can be abused—it can be damaged in a wasteful way. But again, there's no particular reason it should be programmed to mind such treatment" (ibid., 72). Understood in this way, computers, robots, and other mechanisms are situated outside the scope of moral consideration or "beyond good and evil" (Nietzsche 1966, 206). As such, they cannot, strictly speaking, be harmed, nor can or should they be ascribed anything like "rights" that would need to be respected. The only legitimate moral agent is a human programmer or operator, and the only legitimate patient is another human being who would be on the receiving end of any use or application of such technology. Or, to put it another way, because machines have been determined to be nothing but mere instruments of human action, they are neither moral agents (i.e., originators of moral decision and action) nor moral patients (i.e., receivers of moral consideration).

This line of argument, one that obviously draws on and is informed by the instrumentalist definition of technology, is also supported by and mobilized in the field of computer ethics. As Deborah Johnson and Keith Miller (2008) describe it, inadvertently channeling Marshall McLuhan in the process, "computer systems are an extension of human activity" (Johnson and Miller 2008, 127) and "should be understood in ways that keep them conceptually tethered to human agents" (ibid., 131). Following this prosthetic understanding of technology, computer ethics assumes that the information-processing machine, although introducing some new challenges and opportunities for moral decision making and activity, remains a mere instrument or medium of human action. For this reason, the field endeavors to stipulate the appropriate use and/or misuse of technology by human agents for the sake of respecting and protecting the rights of other human patients. In fact, the "Ten Commandments of Computer Ethics," a list first compiled and published by the Computer Ethics Institute (CEI) in 1992, specifies what constitutes appropriate use or misuse of computer technology. The objective of each of the commandments is to stipulate the proper behavior of a human agent for the sake of respecting and protecting the rights of a human patient. "Thou shalt not," the first commandment reads, "use a computer to harm another person." Consequently, computers are, Johnson and Miller (2008, 132) conclude, "deployed by humans, they are used for some human purpose, and they have indirect effects on humans."

On the other hand, the same conceptual arrangement has been employed to argue the exact opposite, namely, that any machine achieving some level of agency would need to be extended consideration of patiency. Indicative of this effort is David Levy's "The Ethical Treatment of Artificially Conscious Robots" (2009) and Robert Sparrow's "The Turing Triage Test" (2004). According to Levy, the new field of roboethics has been mainly interested in questions regarding the effects of robotic decision making and action. "Almost all the discussions within the roboethics community and elsewhere," Levy (2009, 209) writes, "has thus far centered on questions of the form: 'Is it ethical to develop and use robots for such and such a purpose?,' questions based upon doubts about the effect that a particular type of robot is likely to have, both on society in general and on those with whom the robots will interact in particular." Supported by a review of the current literature in the field, Levy argues that roboethics has been exclusively focused on questions regarding both human and machine moral agency. For this reason, he endeavors to turn attention to the question of machine patiency—a question that has, in his estimation, been curiously absent. "What has usually been missing from the debate is the complementary question: 'Is it ethical to treat robots in such-and-such a way?" (ibid.). In taking-up and addressing this other question, Levy refers the matter to the issue of consciousness, which "seems to be widely regarded as the dividing line between being deserving of ethical treatment and not" (ibid., 216). In fact, Levy's investigation, as already announced by its title, is not concerned with the moral status of any and all machines; he is only interested in those that are programmed with "artificial consciousness" (ibid.). "We have," Levy concludes, "introduced the question of how and why robots should be treated ethically. Consciousness or the lack of it has been cited as the quality that generally determines whether or not something is deserving of ethical treatment. Some indications of consciousness have been examined, as have two tests that could be applied to detect whether or not a robot possesses (artificial) consciousness" (ibid., 215). Levy's consideration of machine moral patiency, therefore, is something that is both subsequent and complementary to the question of moral agency. And his argument succeeds or fails, like many of those that have been advanced in investigations of artificial moral agency, on the basis of some test that is able to resolve or at least seriously address the problem of other minds.

A similar maneuver is evident in Sparrow's consideration of AI. "As soon as AIs begin to possess consciousness, desires and projects," Sparrow (2004, 203) suggests, "then it seems as though they deserve some sort of moral standing." In this way, Sparrow, following the reciprocal logic of the "standard position," argues that machines will need to be considered legitimate moral patients the moment that they show recognizable signs of possessing the characteristic markers of agency, which he defines as consciousness, desires, and projects. The question of machine moral patiency, therefore, is referred and subsequent to a demonstration of agency. And for this reason, Sparrow's proposal immediately runs up against an epistemological problem: When would we know whether a machine had achieved the necessary benchmarks for such moral standing? In order to define the ethical tipping point, the point at which a computer becomes a legitimate subject of moral concern, Sparrow proposes, as Allen, Varner, and Zinser (2000) had previously done, a modification of the Turing test. Sparrow's test, however, is a bit different. Instead of determining whether a machine is capable of passing as a human moral agent, Sparrow's test asks "when a computer might fill the role of a human being in a moral dilemma" (Sparrow 2004, 204). The dilemma in question is the case of medical triage, literally a life-and-death decision concerning two different forms of patients.

In the scenario I propose, a hospital administrator is faced with the decision as to which of two patients on life support systems to continue to provide electricity to, following a catastrophic loss of power in the hospital. She can only preserve the existence of one and there are no other lives riding on the decision. We will know that machines have achieved moral standing comparable to a human when the replacement of one of the patients with an artificial intelligence leaves the character of the dilemma intact. That is, when we might sometimes judge that it is reasonable to preserve the continued existence of the machine over the life of the human being. This is the "*Turing Triage Test.*" (Sparrow 2004, 204)

As it is described by Sparrow, the Turing triage test evaluates whether and to what extent the continued existence of an AI may be considered to be comparable to another human being in what is arguably a highly constrained and somewhat artificial situation of life and death. In other words, it may be said that an AI has achieved a level of moral standing that is at least on par with that of another human being, when it is possible that one could in fact choose the continued existence of the AI over that of another human individual—or, to put it another way, when the human and AI system have an equal and effectively indistinguishable "right to life." This decision, as Sparrow points out, would need to be based on the perceived "moral status" of the machine and the extent to which one was convinced it had achieved a level of "conscious life" that would be equivalent to a human being. Consequently, even when the machine is a possible candidate of moral concern, its inclusion in the community of moral patients is based on and derived from a prior determination of agency.

What is interesting, however, is not the different ways the "standard position" comes to be used. What is significant is the fact that this line of reasoning has been, at least in practice, less than standard. In fact, the vast majority of research in the field, as Levy's (2009) literature review indicates, extends consideration of moral agency to machines but gives little or no serious thought to the complementary question of machine moral patiency. Despite the fact that Luciano Floridi and J. W. Sanders (2004) designate this "non-standard," it comprises one of the more common and accepted approaches. To further complicate things, Floridi (1999, 42) integrates these various agent-oriented approaches under the umbrella term "standard" or "classic" in distinction to a "patient-oriented ethics," which he then calls "non-standard." Consequently, there are two seemingly incompatible senses in which Floridi employs the terms "standard" and "non-standard." On the one hand, the "standard position" in ethics "maintains that all entities that qualify as moral agents also qualify as moral patients" (Floridi and Sanders 2004, 350). Understood in this fashion, "non-standard" indicates any asymmetrical and unequal relationship between moral agent and patient. On the other hand, "standard" refers to the anthropocentric tradition in ethics that is exclusively agent-oriented, no matter the wide range of incompatibilities that exist, for example, between virtue ethics, consequentialism, and deontologism. Understood in this fashion, "non-standard" would indicate any ethical theory that was patient-oriented.

Consequently, even if the majority of research in and published work on machine morality is "non-standard" in the first sense, that is, asymmetrically agent-oriented, it is "standard" in the second sense insofar as it is arranged according to the agent-oriented approach developed in and favored by the history of moral philosophy. As J. Storrs Hall (2001, 2) insightfully points out, "we have never considered ourselves to have moral duties to our machines," even though we have, as evidenced by the previous chapter, spilled a considerable amount of ink on the question of whether and to what extent machines might have moral duties and responsibilities to us. This asymmetry becomes manifest either in a complete lack of consideration of the machine as moral patient or by the fact that the possibility of machine moral patiency is explicitly identified as something that is to be excluded, set aside, or deferred.

The former approach is evident in recently published journal articles and conference papers that give exclusive consideration to the issue of machine moral agency. Such exclusivity is announced and immediately apparent in titles such as "On the Morality of Artificial Agents" (Floridi and Sanders 2004), "When Is a Robot a Moral Agent?" (Sullins 2006),

"Ethics and Consciousness in Artificial Agents" (Torrance 2008), "Prolegomena to Any Future Artificial Moral Agent" (Allen, Varner, and Zinser 2000), "The Ethics of Designing Artificial Agents" (Grodzinsky, Miller, and Wolf 2008), "Android Arete: Toward a Virtue Ethic for Computational Agents" (Coleman 2001), "Artificial Agency, Consciousness, and the Criteria for Moral Agency: What Properties Must an Artificial Agent Have to Be a Moral Agent?" (Himma 2009), "Information, Ethics, and Computers: The Problem of Autonomous Moral Agents" (Stahl 2004). These texts, as their titles indicate, give detailed consideration to the question of machine moral agency. They do not, however, reciprocate and give equal attention to the question of machine moral patiency. This lack or absence, however, is never identified or indicated as such. It only becomes evident insofar as these investigations already deviate from the reciprocity stipulated and predicted by the "standard position." In other words, it is only from a perspective informed by the "standard position," at least as it is defined by Floridi and Sanders (2004), that this lack of concern with the complementary question of patiency becomes evident and identifiable. These documents, one can say, literally have nothing to say about the question of the machine as a moral patient.

This does not, however, imply that such investigations completely ignore or avoid the question of moral patiency altogether. As Thomas McPherson (1984, 173) points out, "the notion of a moral agent generally involves that of a patient. If someone performs an act of torture, somebody else must be tortured; if someone makes a promise, he must make it to someone, etc. The notion of a moral agent makes no sense in total isolation from that of the patient." The various publications addressing machine moral agency, therefore, do not simply ignore the question of patiency tout court, which would be logically inconsistent and unworkable. They do, however, typically restrict the population of legitimate moral patients to human beings and human institutions. In fact, the stated objective of these research endeavors, the entire reason for engaging in the question of machine moral agency in the first place, is to investigate the impact autonomous machines might have on human assets and interests. As John Sullins (2006, 24) aptly describes it, "a subtle, but far more personal, revolution has begun in home automation as robot vacuums and toys are becoming more common in homes around the world. As these machines increase in capacity and ubiquity, it is inevitable that they will impact our lives ethically as well as physically and emotionally. These impacts will be both positive and negative and in this paper I will address the moral status of robots and how that status, both real and potential, should affect the way we design and use these technologies." For this reason, these supposed innovations turn out to be only half a revolution in moral thinking; they contemplate extending moral agency beyond the traditional boundaries of the human subject, but they do not ever give serious consideration for doing the same with regard to moral patiency.

Not every examination, however, deploys this distinctly agent-oriented approach without remark, recognition, or reflection. There are a few notable exceptions—notable because they not only make explicit reference to the problem of machine moral patiency but also, and despite such indications, still manage to exclude the machine from the rank and file of legitimate moral subjects. This maneuver is evident, for example, in the project of machine ethics (ME). Like many publications addressing machine moral agency, ME is principally concerned with autonomous machine decision making and responsibility. But unlike many of the texts addressing this subject matter, it marks this exclusive decision explicitly and right at the beginning. "Past research concerning the relationship between technology and ethics has largely focused on responsible and irresponsible use of technology by human beings, with a few people being interested in how human beings ought to treat machines" (Anderson, Anderson, and Armen 2004, 1). In this, the first sentence of the first paper addressing the project of ME, the authors Michael Anderson, Susan Leigh Anderson, and Chris Armen begin by distinguishing their approach from two others. The first is computer ethics, which is concerned, as Anderson and company correctly point out, with questions of human action through the instrumentality of computers and related information systems. In clear distinction from these efforts, machine ethics seeks to enlarge the scope of moral agents by considering the ethical status and actions of machines. As Anderson and Anderson (2007a, 15) describe it in a subsequent publication, "the ultimate goal of machine ethics, we believe, is to create a machine that *itself* follows an ideal ethical principle or set of principles."

The other exclusion addresses the machine as moral patient, or "how human beings ought to treat machines." This also does not fall under the purview of ME, and Anderson, Anderson, and Armen explicitly mark it as something to be set aside by their own endeavors. Although the "question of whether intelligent machines should have moral standing," Susan Leigh Anderson (2008, 480) writes in another article, appears to "loom on the horizon," ME pushes this issue to the margins. This means, then, that ME only goes halfway in challenging the "human-centered perspective" (Anderson, Anderson, and Armen 2004, 1) that it endeavors to address and remediate. ME purports to question the anthropocentrism of moral agency, providing for a more comprehensive conceptualization that can take intelligent and/or autonomous machines into account. But when it comes to moral patiency, only human beings constitute a legitimate subject. In fact, ME is primarily and exclusively concerned with protecting human assets from potentially dangerous machine decisions and actions (Anderson and Anderson 2007a). For this reason, ME does not go very far in questioning the inherent anthropocentrism of the moral patient. In fact, it could be said that considered from the perspective of the patient, ME reasserts the privilege of the human and considers the machine only insofar as we seek to protect the integrity and interests of the human being. Although significantly expanding the subject of ethics by incorporating the subjectivity and agency of machines, ME unfortunately does not provide for a serious consideration of the response to and responsibility for these ethical programmed mechanisms. Such an ethics, despite considerable promise and explicit declarations to the contrary, retains a distinctly "human-centered perspective."

A similar kind of dismissal is operationalized in the work of J. Storrs Hall. Hall's efforts are significant, because he is recognized as one of the first AI researchers to take up and explicitly address the machine question in ethics. His influential article, "Ethics for Machines" (2001), which Michael Anderson credits as having first introduced and formulated the term "machine ethics," describes the problem succinctly:

Up to now, we haven't had, or really needed, similar advances in "ethical instrumentation." The terms of the subject haven't changed. Morality rests on human shoulders, and if machines changed the ease with which things were done, they did not change the responsibilities for doing them. People have always been the only "moral agents." Similarly, people are largely the objects of responsibility. There is a developing debate over our responsibilities to other living creatures, or species of them. . . . We have never, however, considered ourselves to have "moral" duties to our machines, or them to us. (Hall 2001, 2)

Despite this statement, which explicitly recognizes the exclusion of the machine from the ranks of both moral agency and patiency, Hall's work proceeds to give exclusive attention to the former. Like the project of machine ethics, the primary focus of Hall's "Ethics for Machines" is on protecting human assets and interests from potentially dangerous machine actions and decision making. "We will," Hall (2001, 6) predicts, "all too soon be the lower-order creatures. It will behoove us to have taught them (intelligent robots and AIs) well their responsibilities toward us."

This exclusive focus on machine moral agency persists in Hall's subsequent book-length analysis, *Beyond AI: Creating the Conscience of the Machine* (2007). Although the term "artificial moral agency" occurs throughout the text, almost nothing is written about the possibility of "artificial moral patiency," which is a term Hall does not consider or utilize. The closest *Beyond AI* comes to addressing the question of machine moral patiency is in a brief comment situated in the penultimate chapter, "The Age of Virtuous Machines." "Moral agency," Hall (2007, 349) writes, "breaks down into two parts—rights and responsibilities—but they are not coextensive. Consider babies: we accord them rights but not responsibilities. Robots are likely to start on the other side of that inequality, having responsibilities but not rights, but, like babies, as they grow toward (and beyond) full human capacity, they will aspire to both." This statement is remarkable for at least two reasons. First, it combines what philosophers have typically distinguished as moral agent and patient into two aspects of agency—responsibilities and rights. This is, however, not simply a mistake or slip in logic. It is motivated by the assumption, as Hajdin (1994) pointed out, that moral patiency is always and already derived from and dependent on the concept of agency. In the conceptual pair agent–patient, "agent" is the privileged term, and patiency is something that is derived from it as its opposite and counterpart. Even though Hall does not use the term "artificial moral patient," it is already implied and operationalized in the concept "moral rights."

Second, formulated in this way, a moral agent would have both responsibilities and rights. That is, he/she/it would be both a moral agent, capable of acting in an ethically responsible manner, and a moral patient, capable of being the subject of the actions of others. This kind of symmetry, however, does not necessarily apply to all entities. Human babies, Hall points out (leveraging one of the common examples), would be moral patients well in advance of ever being considered legitimate moral agents. Analogously, Hall suggests, AI's and robots would first be morally responsible agents prior to their ever being considered to have legitimate claims to moral rights. For Hall, then, the question of "artificial moral agency" is paramount. The question of rights or "artificial moral patiency," although looming on the horizon, as Susan Leigh Anderson (2008, 480) puts it, is something that is deferred, postponed, and effectively marginalized.

A similar decision is deployed in Wendell Wallach and Colin Allen's *Moral Machines* (2009), and is clearly evident in the choice of the term "artificial moral agent," or AMA, as the protagonist of the analysis. This term, which it should be mentioned had already been utilized well in advance of Hall's *Beyond AI* (see Allen, Varner, and Zinser 2000), immediately focuses attention on the question of agency. Despite this exclusive concern, however, Wallach and Allen (2009, 204–207) do eventually give brief consideration not only to the legal responsibilities but also the rights of machines. Extending the concept of legal responsibility to AMAs is, in Wallach and Allen's opinion, something of a no–brainer: "The question whether there are barriers to designating intelligent systems legally accountable for their actions has captured the attention of a small but growing community of scholars. They generally concur that the law, as it exists can accommodate the advent of intelligent (ro)bots. A vast body of law already exists for attributing legal personhood to nonhuman entities (corporations). No radical changes in the law would be required to extend the status of legal person to machines with higher-order faculties, presuming that the (ro)bots were recognized as responsible agents" (ibid., 204). According to Wallach and Allen's estimations, a decision concerning the legal status of AMA's should not pose any significant problems. Most scholars, they argue, already recognize that this is adequately anticipated by and already has a suitable precedent in available legal and judicial practices, especially as it relates to the corporation.

What is a problem, in their eyes, is the flip side of legal responsibility— the question of rights. "From a legal standpoint," Wallach and Allen continue, "the more difficult question concerns the rights that might be conferred on an intelligent system. When or if future artificial moral agents should acquire legal status of any kind, the question of their legal rights will also arise" (ibid.). Although noting the possibility and importance of the question, at least as it would be characterized in legal terms, they do not pursue its consequences very far. In fact, they mention it only to defer it to another kind of question—a kind of investigative bait and switch: "Whether or not the legal ins and outs of personhood can be sorted out, more immediate and practical for engineers and regulators is the need to evaluate AMA performance" (ibid., 206). Consequently, Wallach and Allen conclude *Moral Machines* by briefly gesturing in the direction of a consideration of the machine as moral patient only to refer this question back to the issue of machine agency and performance measurement. In this way, then, Wallach and Allen briefly move in the direction of the question of patiency only to immediately recoil from the complications it entails, namely, the persistent philosophical problem of having to sort out the ins and outs of moral personhood. Although not simply passing over the question of machine moral patiency in silence, Wallach and Allen, like Anderson et al. and Hall, only mention it in order to postpone or otherwise exclude the issue from further consideration.

If one were generous in his or her reading, it might be possible to excuse such exclusions and deferrals as either a kind of momentary oversight or the unintended by-product of a focused investigative strategy. These texts, it could be argued, are not intended to be complete philosophical investigations of all aspects of the machine question. They are, more often than not, simply exercises in applied moral philosophy that endeavor to address very specific problems in the design, programming, and deployment of artificial autonomous agents. Despite these excuses, however, such efforts do have significant metaphysical and moral consequences. First, an exclusive concern with the question of machine moral agency, to the almost absolute omission of any serious consideration of patiency, comprises a nonstandard, asymmetrical moral position that Floridi and Sanders (2004, 350) term "unrealistic." "This pure agent," they note, "would be some sort of supernatural entity that, like Aristotle's God, affects the world but can never be affected by it" (ibid., 377). According to Floridi and Sanders, then, any investigative effort that, for whatever reason, restricts the machine to questions of moral agency without consideration of its reciprocal role as a legitimate patient, has the effect, whether intended or not, of situating the machine in a position that has been and can only be occupied by a supernatural entity—the very *deus ex machina* of science fiction. For this reason, as Floridi and Sanders conclude, "it is not surprising that most macroethics have kept away from these 'supernatural' speculations" (ibid.). Although the various texts addressing machine moral agency appear to be rather sober, pragmatic, and empirical in their approach, they already deploy and depend on a metaphysical figure of "pure agency" that is both unrealistic and speculative.

Second, this concept of "pure agency" has considerable ethical complications. It is, as Kari Gwen Coleman (2001, 253) recognizes, a "slave ethic," where "the computational agents under consideration are essentially slaves whose interests—if they can be said to have them—are just those of the humans whom they serve." The axiological difficulties associated with this kind of moral stance are often illustrated by way of Isaac Asimov's three laws of robotics, addressed in the previous chapter. Articulated in the form of three imperatives stipulating proper robotic behavior, Asimov's laws, which have had considerable influence in discussions of AI, robotics, and ethics (Anderson 2008), provide explicit recognition of robots as morally accountable agents. In doing so, Asimov's fictional stories advance one step further than computer ethics, which simply and immediately dismisses the machine from any consideration of moral accountability or responsibility. Despite this apparent advance, however, the letter of the laws indicates little or nothing concerning the machine as a moral patient. In other words, the laws stipulate how robots are to respond to and interact with human beings but say nothing, save the third law's stipulation of a basic right to continued existence, concerning any responsibilities that human users might have to such ethically minded or programmed machines. And it is precisely this aspect of the three laws that has been the target of critical commentary. According to Aaron Sloman's (2010, 309) reading, "Asimov's laws of robotics are immoral, because they are unfair to future robots which may have their own preferences, desires and values." Sloman's criticism, which appeals to a sense of equal treatment and reciprocity, leverages Floridi and Sanders's (2004) "standard position" to argue that anyone or anything that is accorded the status of moral agency must also be considered a moral patient. Following this assumption, Sloman concludes that any effort to impose stipulations of moral agency on robots or intelligent machines without also taking into account aspects of their legitimate claim to moral patiency would be both unjustified and immoral.

This interpretation of Asimov's laws, however, is incomplete and not entirely attentive to the way the laws have been developed and come to be utilized in his stories. If one only reads the letter of the laws, it may be accurate to conclude that they provide little or no consideration of machine moral patiency. As we saw in the last chapter, Asimov introduced the laws not as some complete moral code for future robotic entities but as a literary device for generating fictional stories. The ensuing narratives, in fact, are often about the problems caused by the laws, especially as they relate to robot rights, legal status, and questions of moral patiency. The short story "The Bicentennial Man" (Asimov 1976), for instance, begins with a restatement of the three laws and narrates the experiences of a robot named Andrew, who was programmed to operate within the parameters they stipulate. The plot of the story concerns Andrew's development and his struggle to be granted basic "human rights." "The Bicentennial Man," therefore, is motivated by and investigates the problems of stipulating moral agency without also giving proper consideration to the question and possibility of machine moral patiency. As Susan Leigh Anderson (2008, 484) writes in her critical reading of the story, "if the machine is given principles to follow to guide its own behavior . . . an assumption must be made about its status. The reason for this is that in following any ethical theory the agent must consider at least him/her/itself, if he/she/it has moral standing, and typically others as well, in deciding how to act. As a result, a machine agent must know if it is to count, or whether it must always defer to others who count while it does not, in calculating the correct action in a moral dilemma." What Asimov's story illustrates, therefore, is the problem of stipulating a code of behavior without also giving serious consideration to questions of moral patiency. To put it another way, the three laws intentionally advance a nonstandard ethical position, one that deliberately excludes considerations of patiency, in order to generate stories out of the conflict that this position has with the standard moral position.

As long as moral patiency is characterized and conceptualized as nothing other than the converse and flip side of moral agency, it will remain secondary and derivative. What is perhaps worse, this predominantly agent-oriented approach comprises what Friedrich Nietzsche (1966, 204) had termed a "master morality," whereby membership in the community of moral subjects would be restricted to one's peers and everything else would be excluded as mere objects to be used and even abused without any axiological consideration whatsoever. "A morality of the ruling group," Nietzsche writes, "is most alien and embarrassing to the present taste in the severity of its principle that one has duties only to one's peers; that against beings of a lower rank, against everything alien, one may behave as one pleases or 'as the heart desires,' and in any case 'beyond good and evil'" (ibid., 206). Perhaps one of the best illustrations of this can be found in Homer's *Odyssey*. "When god-like Odysseus," Aldo Leopold (1966, 237) recalls, "returned from the wars in Troy, he hanged all on one rope a dozen slave-girls of his household whom he suspected of misbehavior during his absence. This hanging involved no question of propriety. The girls were property. The disposal of property was then, as now, a matter of expediency, not of right and wrong." As long as others—whether human, animal, machine, or otherwise—are defined as mere instruments or the property of a ruling group, they can justifiably be used, exploited, and dispensed with in a way that is purely expedient and beyond any moral consideration whatsoever.

In response to these perceived difficulties, philosophers have recently sought to articulate alternative concepts of moral patiency that break with or at least significantly complicate this precedent. These innovations deliberately invert the agent-oriented approach that has been the standard operating presumption of moral philosophy and institute a "patient-oriented ethics," as Floridi (1999, 42) calls it, that focuses attention not on the perpetrator of an act but on the victim or receiver of the action. For this reason, this alternative is often called "nonstandard" or "nonclassic" in order to differentiate it from the traditional forms of agent-oriented moral thinking. As Floridi neatly characterizes it, "classic ethics are philosophies of the wrongdoer, whereas non-classic ethics are philosophies of the victim. They place the 'receiver' of the action at the center of the ethical discourse, and displace its 'transmitter' to its periphery" (ibid.). Although there is as yet little research in the application of this nonstandard, patient-oriented approach to autonomous machines, two recent innovations hold considerable promise for this kind of patient-oriented approach to moral thinking—animal ethics and information ethics.

2.3 The Question of the Animal

Traditional forms of agent-oriented ethics, no matter how they have come to be articulated (e.g., virtue ethics, utilitarian ethics, deontological ethics), have been anthropocentric. This has the effect (whether intended or not) of excluding others from the domain of ethics, and what gets left out are, not surprisingly, nonhuman animals and their Cartesian counterpart, machines. It is only recently that the discipline of philosophy has begun to approach nonhuman animals as a legitimate subject of ethics. According to Cary Wolfe (2003a,b), there are two factors that motivated this remarkable reversal of the anthropocentric tradition. On the one hand, there is the crisis of humanism, "brought on, in no small part, first by structuralism and then poststructuralism and its interrogation of the figure of the human as the constitutive (rather than technically, materially, and discursively constituted) stuff of history and the social" (Wolfe 2003a, x–xi). Since at least Nietzsche, philosophers, anthropologists, and social scientists have been increasingly suspicious of the privileged position human beings have given themselves in the great chain of being, and this suspicion has become an explicit object of inquiry within the so-called human sciences.

On the other hand, the boundary between the animal and the human has, as Donna Haraway (1991, 151–152) remarks, become increasingly untenable. Everything that had divided us from them is now up for grabs: language, tool use, and even reason. Recent discoveries in various branches of the biological sciences have had the effect of slowly dismantling the wall that Descartes and others had erected between the human and the animal other. According to Wolfe (2003a, xi), "a veritable explosion of work in areas such as cognitive ethology and field ecology has called into question our ability to use the old saws of anthropocentrism (language, tool use, the inheritance of cultural behaviors, and so on) to separate ourselves once and for all from the animals, as experiments in language and cognition with great apes and marine mammals, and field studies of extremely complex social and cultural behaviors in wild animals such as apes, wolves, and elephants, have more or less permanently eroded the tidy divisions between human and nonhuman." The revolutionary effect of this transformation can be seen, somewhat ironically, in the backlash of what Evan Ratliff (2004) calls "creationism 2.0," a well-organized "crusade against evolution" that attempts to reinstate a clear and undisputed division between human beings and the rest of animal life based on a strict interpretation of the Judeo-Christian creation myth. What is curious in this recent questioning and repositioning of the animal is that its other, the machine, remains conspicuously absent. Despite all the talk of the animal question, animal others, animal rights, and the reconsideration of what Wolfe (2003a, x) calls the "repressed Other of the subject, identity, logos," little or nothing has been said about the machine.

Despite this exclusion, a few researchers and scholars have endeavored to connect the dots between animal ethics and the machine. David Calverley, for example, has suggested that animal rights philosophy provides an opportunity to consider machines as similarly situated moral patients:

As a result of modern science, animals have been shown to possess, to varying degrees, characteristics that, taken in the aggregate, make them something more than inanimate objects like rocks but less than human. These characteristics, to the extent that they are a valid basis for us to assert that animals have a claim to moral consideration, are similar to characteristics designers are seeking to instantiate in androids. If the designers succeed with the task they have set for themselves, then logically androids, or someone acting on their behalf in some form of guardianship relationship, could assert claims to moral consideration in a manner similar to those claimed for animals. (Calverley 2006, 408)

Unlike Descartes, however, Calverley does not simply assert the connection as a matter of fact but advocates that we "examine both the similarities and the differences between the two in some detail to test the validity of the analogy" (ibid.). The crucial issue, therefore, is to determine, as David Levy (2009) points out in response to Calverley's argument, to what extent the analogy holds. If, for example, we can demonstrate something approaching the Cartesian level of association between animals and machines, or even some limited analogical interaction between the two, then the extension of moral rights to animals would, in order to be both logically and morally consistent, need to take seriously the machine as a similar kind of moral patient. If, however, important and fundamental differences exist that would permit one to distinguish animals from machines, then one will need to define what these differences are and how they determine and justify what is and what is not legitimately included in the community of morally significant subjects.

So let's start at the beginning. What is now called "animal rights philosophy," as Peter Singer points out, has a rather curious and unlikely origin story:

The idea of "The Rights of Animals" actually was once used to parody the case for women's rights. When Mary Wollstonecraft, a forerunner of today's feminists [and also the mother of Mary Shelley], published her *Vindication of the Rights of Women* in 1792, her views were regarded as absurd, and before long an anonymous publication appeared entitled *A Vindication of the Rights of Brutes*. The author of this satirical work (now known to have been Thomas Taylor, a distinguished Cambridge philosopher) tried to refute Mary Wollstonecraft's arguments by showing how they could be carried one stage further. (Singer 1975, 1)

The discourse of animal right, then, begins as parody. It was advanced as a kind of reductio ad absurdum in order to demonstrate the conceptual failings of Wollstonecraft's proto-feminist manifesto. The argument utilizes, derives from, and in the process makes evident a widely held assumption that has, for a good part of the history of moral philosophy, gone largely uninvestigated—that women, like animals, have been excluded from the subject of moral reasoning. As Matthew Calarco describes it by way of an analysis of Derrida's writings on the animal:

the meaning of subjectivity is constituted through a network of exclusionary relations that goes well beyond a generic human–animal distinction . . . the metaphysics of subjectivity works to exclude not just animals from the status of being full subjects but other beings as well, in particular women, children, various minority groups, and other Others who are taken to be lacking in one or another of the basic traits of subjectivity. Just as many animals have and continue to be excluded from basic legal protections, so, as Derrida notes, there have been "many 'subjects' among mankind who are not recognized as subjects" and who receive the same kind of violence typically directed at animals. (Calarco 2008, 131)

In other words, Taylor's parody leveraged and was supported by an assumption that women, like animals, have often been excluded from being full participants in moral considerations. For this reason, making a case for the "vindication of the rights of women" would, in Taylor's estimations, be tantamount to suggesting the same for "brutes."

For Singer, however, what began as parody turns out to be a serious moral issue. And this is, according to Singer's account of the genealogy, taken up and given what is perhaps its most emphatic articulation in Jeremy Bentham's *An Introduction to the Principles of Morals and Legislation*. For Bentham, the question of ethical treatment did not necessarily rest on the notion of some shared sense of rationality. Even if it could be shown that a horse or dog had more reason than a human infant, the faculty of reason was not determinative. "The question," Bentham (2005, 283) wrote, "is not, Can they reason? nor Can they talk? but, Can they suffer?" Following this change in the fundamental moral question, Singer (1975, 8) argues that it is "the capacity for suffering" or more strictly defined "the capacity for suffering and/or enjoyment or happiness" that should determine what is and what is not included in moral considerations. "A stone," Singer argues, "does not have interests because it cannot suffer. Nothing that we can do to it could possibly make any difference to its welfare. A mouse, on the other hand, does have an interest in not being kicked along the road, because it will suffer if it is" (ibid., 9). The issue of suffering, then, has the effect, Derrida (2008, 27) points out, of "changing the very form of the question regarding the animal":

> Thus the question will not be to know whether animals are of the type *zoon logon echon* [ζῷον λόγον ἔχον] whether they *can* speak or reason thanks to that *capacity* or that *attribute* of the *logos* [λόγος], the *can-have* of the *logos*, the aptitude for the *logos* (and logocentrism is first of all a thesis regarding the animal, the animal deprived of the *logos*, deprived of the *can-have-the-logos*: this is the thesis, position, or presumption maintained from Aristotle to Heidegger, from Descartes to Kant, Levinas, and Lacan). The *first* and *decisive* question would be rather to know whether animals *can suffer*. (Ibid.)

The shift, then, is from the possession of a certain ability or power to do something (λόγος) to a certain passivity—the vulnerability of not-being-able. Although Derrida and Singer do not use the term, this is a patient-oriented approach to ethics that does not rely on moral agency or its qualifying characteristics (e.g., reason, consciousness, rationality, language). The main and only qualifying question is "can they suffer," and this has to do with a certain passivity—the patience of the patient, words that are derived from the Latin verb *patior*, which connotes "suffering." It is, on this view, the common capacity for suffering that defines who or what comes to be included in the moral community.

> If a being suffers there can be no moral justification for refusing to take that suffering into consideration. No matter what the nature of the being, the principle of equality requires that its suffering be counted equally with the like suffering—in so far as rough comparisons can be made—of any other being. If a being is not capable of suffering, or of experiencing enjoyment or happiness, there is nothing to be taken into account. So the limit of sentience (using the term as a convenient shorthand for the capacity to suffer and/or experience enjoyment) is the only defensible boundary of concern for the interests of others. To mark this boundary by some other characteristic like intelligence or rationality would be to mark it in an arbitrary manner. (Singer 1975, 9)

Thus, according to Singer's argument, the suffering–nonsuffering axis is the only morally defensible and essential point of differentiation. All other divisions—those that have, for example, been determined by intelligence, rationality, or other λόγος based qualities—are arbitrary, inessential, and capricious. According to Singer these are as arbitrary and potentially dangerous as making distinctions based on something as inessential as skin color (ibid.).

This call-to-arms for "animal liberation," as Singer's book is titled, sounds promising. It expands the scope of ethics by opening up consideration to previously excluded others. It takes a patient-oriented approach, where moral duties are defined on the basis of a passive inability, not the presence or lack of a particular ability. And this innovation has subsequently gotten a lot of traction in the fields of moral philosophy and legal studies and in the animal rights movement. Despite this success, however, it may seem unlikely that animal rights philosophy and its focus on the "capacity to suffer" would have anything to contribute to the debate concerning the machine as a similarly constructed moral patient. As John Sullins (2002, 1) has stated, "perhaps one might be able to argue for the ethical status of autonomous machines based on how we treat nonhuman animals. I do not think this is going to be all that fruitful since at best, autonomous machines are a kind of animat, inspired by biology but not partaking in it, and they in no way experience the world as robustly as say a large mammal might."

But this opinion is and remains contentious. It is, in fact, precisely on the basis of "suffering" that the question of moral patiency has been extended, at least in theory, to machines. As Wendell Wallach and Colin Allen (2009, 204) characterize it, "from a legal standpoint, the more difficult question concerns the rights that might be conferred on an intelligent system. When or if future artificial moral agents should acquire legal status of any kind, the question of their legal rights will also arise. This will be particularly an issue if intelligent machines are built with a capacity for emotions of their own, for example the ability to feel pain." In this brief remark, Wallach and Allen presume that the principal reason for extending moral patiency to machines, at least in terms of their legal status, would derive from a capacity for emotion, especially "the ability to feel pain." A machine, in other words, would need to be granted some form of legal rights if it could be harmed or otherwise subjected to adverse stimulus. This assumption, although not explicitly stated as such within the letter of the text, follows the innovations of animal rights philosophy, where the capacity to suffer or feel pain is the defining threshold for determining moral patiency in nonhuman animals. All of this is, of course, situated in the form of a conditional statement: *If* machines are built to feel pain, *then* they will, according to Wallach and Allen, need to be accorded not just moral duties but also moral rights.

A similar maneuver is evident in Robert Sparrow's "Turing Triage Test" (2004, 204), which seeks to decide whether "intelligent computers might achieve the status of moral persons." Following the example provided by Peter Singer, Sparrow first argues that the category "personhood," in this context, must be understood apart from the concept of the human. "Whatever it is that makes human beings morally significant," Sparrow writes, "must be something that could conceivably be possessed by other entities. To restrict personhood to human beings is to commit the error of chauvinism or 'speciesism'" (ibid., 207). Second, this expanded concept of "moral personhood," which is uncoupled from the figure of the human, is in turn minimally defined, again following Singer, "as a capacity to experience pleasure and pain" (ibid.). "The precise description of qualities required for an entity to be a person or an object of moral concern differ from author to author. However it is generally agreed that a capacity to experience pleasure and pain provides a *prima facie* case for moral concern. . . . Unless machines can be said to suffer they cannot be appropriate objects for moral concern at all" (ibid.). As promising as this innovation appears to be, animal rights philosophy has a number of problems both as a patient-oriented ethic in its own right and in its possible extension to considerations of other forms of excluded otherness such as machines.

2.3.1 Terminological Problems

Singer's innovative proposal for a nonanthropocentric, patient-oriented ethics faces at least two problems of terminology. First, Singer does not adequately define and delimit "suffering." According to Adil E. Shamoo and David B. Resnik,

His use of the term "suffering" is somewhat naïve and simplistic. It would appear that Singer uses the term "suffer" as a substitute for "feel pain," but suffering is not the same thing as feeling pain. There are many different types of suffering: unrelieved and uncontrollable pain; discomfort, as well as other unpleasant symptoms, such as nausea, dizziness, and shortness of breath; disability; and emotional distress. However, all of these types of suffering involve much more than the awareness of pain: They also involve self-consciousness, or the awareness that one is aware of something. (Shamoo and Resnik 2009, 220–221)

For Shamoo and Resnik, feeling pain is understood to be significantly different from suffering. Pain, they argue, is simply adverse nerve stimulus. Having pain or being aware of a pain, however, is not sufficient to qualify as suffering. Suffering requires an additional element—consciousness, or the awareness that one is feeling pain. Suffering is, on this account, more than having a pain; it is the recognition that one experiences the pain as pain.

Daniel Dennett (1996, 16–17) makes a similar, although not necessarily identical point, by way of a rather gruesome illustration: "A man's arm has been cut off in a terrible accident, but the surgeons think they can reattach it. While it is lying there, still soft and warm, on the operating table, does it feel pain? A silly suggestion you reply; it takes a mind to feel pain, and as long as the arm is not attached to a body with a mind, whatever you do to the arm can't cause suffering in any mind." For Dennett, it seems entirely possible that an amputated arm, with its network of active nerve cells, does in fact register the adverse stimulus of pain. But in order for that stimulus to be felt as pain, that is, in order for it to be a pain that causes some kind of discomfort or suffering, the arm needs to be attached to a mind, which is presumably where the pain is registered as pain and the suffering takes place.

What these various passages provide, however, is not some incontrovertible and well-established definition of suffering. Rather, what they demonstrate is the persistent and seemingly irreducible terminological slippage associated with this concept. Despite the immediate appearance of something approaching intuitive sense, these various efforts to distinguish pain from suffering remain inconclusive and unsatisfactory. Although Singer, following Bentham's lead, had proposed the criterion "can they suffer" as a replacement for the messy and not entirely accurate concepts of "rationality" and "self-consciousness," suffering easily becomes conflated with and a surrogate for consciousness and mind. Consequently, what had been a promising reconfiguration of the entire problem becomes more of the same.

Second, and directly following from this, Singer's text conflates suffering and sentience. The identification of these two terms is marked and justified in a brief parenthetical aside: "sentience (using the term as a convenient if not strictly accurate shorthand for the capacity to suffer and/or experience enjoyment) is the only defensible boundary of concern for the interests of others" (Singer 1975, 9). For Singer, then, "sentience" is roughly defined as "the capacity to suffer and/or experience enjoyment." Or, as Steve Torrance (2008, 503) describes it, "the notion of sentience should be distinguished from that of self-consciousness: many beings, which possess the former may not possess the latter. Arguably, many mammals possess sentience, or phenomenal consciousness—they are capable of feeling pain, fear, sensuous pleasure and so on." Consequently, Singer's characterization of sentience is less dependent on the Cartesian *cogito ergo sum* and more in line with the material philosophy of the Marquis de Sade, who comprises something of the "dark side" of modern rationalism.

This use of the term "sentience," however, may not be, as Singer explicitly recognizes, entirely accurate or strictly formulated. Despite the fact that, as Dennett correctly points out, "there is no established meaning to the word 'sentience'" (Dennett 1996, 66), "everybody agrees that sentience requires sensitivity plus some further as yet unidentified factor x" (ibid., 65). Although there is considerable debate in the philosophy of mind, neuroscience, and bioethics as to what this "factor x" might be, the fact of the matter is that defining sentience as "the capability to suffer" runs the risk of undermining Bentham's initial moral innovation. As Derrida explains, Bentham's question is a radical game changer:

"Can they suffer?" asks Bentham, simply yet so profoundly. Once its protocol is established, the form of this question changes everything. It no longer simply concerns the *logos*, the disposition and whole configuration of the *logos*, having it or not, nor does it concern, more radically, a *dynamis* or *hexis*, this having or manner of being, this *habitus* that one calls a faculty or "capability," this can-have or the power one possesses (as in the power to reason, to speak, and everything that that implies). The question is disturbed by a certain *passivity*. It bears witness, manifesting already, as question, the response that testifies to a sufferance, a passion, a not-being-able. (Derrida 2008, 27)

According to Derrida, the question "can they suffer?" structurally resists identification with sentience. In whatever way it comes to be defined, irrespective of what faculty or faculties come to stand in for Dennett's "factor x," sentience is understood to be and is operationalized as an *ability*. That is, it is a power or capacity that one either does or does not possesses— what the ancient Greeks would have characterized as *dynamis* or *hexis*. What makes Bentham's question so important and fundamental, in Derrida's estimation, is that it asks not about an ability of mind (however that would come to be defined) but of a certain *passivity* and irreducible lack.

"'Can they suffer?'" Derrida concludes, "amounts to asking 'Can they *not be able?*'" (ibid., 28). By conflating suffering with sentience, Singer unfortunately and perhaps unwittingly transforms what had been a fundamental form of passivity and patience into a new capability and agency. Interpreted in this fashion, Bentham's question would be reformulated in such a way that it would change little or nothing. Understood as a new capability, the inquiry "Can they suffer?" simply shifts the point of comparison by lowering the level of abstraction. In this way, the qualifying criterion for membership in the moral community would no longer be the capacity for reason or speech but the ability to experience pain or pleasure. This domestication of Bentham's potentially radical question achieves its natural endpoint in *The Case for Animal Rights* (Regan 1983, 2), in which Tom Regan affirms and argues for the attribution of consciousness and a mental life to animals. Once consciousness—no matter how it is defined or characterized—enters the mix, we are returned to the fundamental epistemological question that had caused significant difficulties for the consideration of moral agency: If animals (or machines) have an inner mental life, how would we ever know it?

2.3.2 Epistemological Problems

Animal rights philosophy, following Bentham, changes the operative question for deciding moral standing and who or what comes to be included in the community of moral subjects. The way this question has been taken up and investigated, however, does not necessarily escape the fundamental epistemological problem. As Matthew Calarco (2008, 119) describes it, the principal concern of animal rights philosophy, as developed in the Anglo-American philosophical tradition, has "led to an entire field of inquiry focused on determining whether animals actually suffer and to what extent this can be confirmed empirically." Whether the qualifying criterion is the capacity for λόγος (characterized in terms like consciousness, intelligence, language, etc.) or the capability to suffer (what Singer designates with the word "sentience"), researchers are still confronted with a variant of the other minds problem. How, for example, can one know that an animal or even another person actually suffers? How is it possible to access and evaluate the suffering that is experienced by another? "Modern philosophy," Calarco writes, "true to its Cartesian and scientific aspirations, is interested in the indubitable rather than the undeniable. Philosophers want proof that animals actually suffer, that animals are aware of their suffering, and they require an argument for why animal suffering should count on equal par with human suffering" (ibid.). But such indubitable and certain knowledge appears to be unattainable:

At first sight, "suffering" and "scientific" are not terms that can or should be considered together. When applied to ourselves, "suffering" refers to the subjective experience of unpleasant emotions such as fear, pain and frustration that are private and known only to the person experiencing them (Blackmore 2003, Koch 2004). To use the term in relation to non-human animals, therefore, is to make the assumption that they too have subjective experiences that are private to them and therefore unknowable by us. "Scientific" on the other hand, means the acquisition of knowledge through the testing of hypotheses using publicly observable events. The problem is that we know so little about human consciousness (Koch 2004) that we do not know what publicly observable events to look for in ourselves, let alone other species, to ascertain whether they are subjectively experiencing anything like our suffering (Dawkins 2001, M. Bateson 2004, P. Batson 2004). The scientific study of animal suffering would, therefore, seem to rest on an inherent contradiction: it requires the testing of the untestable. (Dawkins 2008, 1)

Because suffering is understood to be a subjective and private experience, there is no way to know, with any certainty or credible empirical method, how another entity experiences unpleasant emotions such as fear, pain, or frustration. For this reason, it appears that the suffering of another—especially an animal—remains fundamentally inaccessible and unknowable. As Singer (1975, 11) readily admits, "we cannot directly experience anyone else's pain, whether that 'anyone' is our best friend or a stray dog. Pain is a state of consciousness, a 'mental event,' and as such it can never be observed."

A similar difficulty is often recorded when considering machines, especially machines programmed to manifest what appear to be emotional responses. In *2001: A Space Odyssey*, for example, Dave Bowman is asked whether HAL, the shipboard computer, has emotions. In response, Bowman answers that HAL certainly acts as if he has "genuine emotions," but admits that it is impossible to determine whether these are in fact "real feelings" or just clever programming tricks designed into the AI's user interface. The issue, therefore, is how to decide whether the appearance of emotion is in fact the product of real feeling or just an external manifestation and simulation of emotion. This is, as Thomas M. Georges (2003, 108) points out, another version of the question "can machines think?" which inevitably runs up against the epistemological problem of other minds. As Georges explains, connecting the conceptual dots between machines and animals, "people are beginning to accept the idea of a machine that displays the outward appearance of being happy, sad, puzzled, or angry or responds to stimuli in various ways, but they say this is just window dressing. The simulation is transparent in the case of a Happy Face displayed on a monitor screen. We do not mistake it for real feelings any more than we would the smile of a teddy bear. But as emulations get better and better, when might we say that anything resembling human emotions is actually going on inside among the gears, motors, and integrated circuits? And what about nonhuman animals? Do they have emotions?" (ibid., 107–108).

This epistemological limitation does not, at least on Singer's account, foreclose inquiry. Even though we cannot ever get inside the head of another person or animal to know exactly whether and how they experience pain or any other emotion, we can, Singer (1975, 11) argues, "infer that others are feeling it from various external indications." Singer demonstrates this point by redeploying a version of the Cartesian automaton hypothesis:

In theory, we *could* always be mistaken when we assume that other human beings feel pain. It is conceivable that our best friend is a very cleverly constructed robot, controlled by a brilliant scientist so as to give all the signs of feeling pain, but really no more sensitive than any other machine. We can never know, with absolute certainty, that this is not the case. But while this might present a puzzle for philosophers, none of us has the slightest real doubt that our best friends feel pain just as we do. This is an inference, but a perfectly reasonable one based on observations of their behavior in situations in which we would feel pain, and on the fact that we have every reason to assume that our friends are beings like us, with nervous systems like our own that can be assumed to function as ours do, and to produce similar feelings in similar circumstances. If it is justifiable to assume that other humans feel pain as we do, is there any reason why a similar inference should be unjustifiable in the case of other animals? (Ibid., 11–12)

Although seemingly reasonable and grounded in what appears to be common sense, this approach to contending with the problem of other minds—whether human, animal, or machine—has a less than laudable resume. It is, for example, the principal strategy of *physiognomy*, an ancient pseudo-science mistakenly attributed to Aristotle by way of an apocryphal work titled *Physiognomonica*. According to its modern advocate and expositor, Johann Caspar Lavater (1826, 31), "physiognomy is the science or knowledge of the correspondence between the external and internal man, the visible superficies and the invisible contents." This effort to draw formal connections between external bodily expression and internal states of mind, although supported by folk traditions and common assumptions, was widely discredited as "bad science." G. W. F. Hegel, in particular, dedicated a good portion of his *Phenomenology of Spirit* (1801) to a critical assessment of both physiognomy and the related pseudo-science of phrenology. "The 'science of knowing man' [Lavater's term], which deals with the supposed human being, like the 'science' of physiognomy which deals with his presumed reality, and aims at raising the unconscious judging of every day physiognomy to the level of knowledge, is therefore something which lacks both foundation and finality" (Hegel 1977, 193). According to Hegel's analysis, the common practice of physiognomy, no matter how well Lavater or others tried to dress it up in the attire of what might appear to be science, "tells us nothing, that strictly speaking, it is idle chatter, or merely the voicing of one's own opinion" (ibid.). Or as Hegel (1988, 147–148) later summarizes it in the third and final part of the *Encyclopedia of the Philosophical Sciences*, "to try to raise physiognomy . . . to the rank of a science, was therefore one of the vainest fancies, still vainer than a *signature rerum*, which supposed the shape of a plant to afford indication of its medicinal virtue."

Despite being widely discredited as a pseudo-science, the general approach utilized in physiognomy continued to be applied in the more rigorously defined sciences that succeed it. In 1806, for example, Charles Bell published *Anatomy and Philosophy of Expression*, a work that Charles Darwin (1998, 7) argued "laid the foundations of the subject as a branch of science." Darwin, in fact, took up and further developed this science in *The Expression of the Emotions in Man and Animals*. In this work, first published in 1872, Darwin not only examined to what extent different bodily "expressions are characteristic of states of mind" (Darwin 1998, 24) but proposed a principled method for evaluating the emotional state of human beings and animals from the observed physical evidence of their different bodily movements. Although developed in a way that was arguably more scientific than the art of physiognomy, this science also sought to ascertain emotional states from an examination of external expressions—quite literally a "pressing out." Or as Derrida (1973, 32) characterizes it by way of

Edmund Husserl's *Logical Investigations*, "ex-pression is exteriorization. It imparts to a certain outside a sense which is first found in a certain inside."

The main difficulty with these approaches is that they endeavor to make determinations about internal states of mind based on various forms of external evidence. They therefore require something of a "leap of faith," and this problem, as Jennifer Mather (2001, 152) points out, persists in contemporary work in ethology. "Despite not knowing what they might feel, it is relatively easy for me to take a leap of faith and recognize the dog who cringes before punishment, the cats who scream in pain when their paws are crushed, and assume that they are in pain or suffering. It is much less easy for me to decide that one of my octopuses who recoils from contact with a sea anemone is hurting or that a lobster feels pain when being boiled." The problem with relying on inferences and assumptions based on what Singer (2000, 36) calls "various external indicators" is that it always requires "a leap of faith" that is neither rigorously applied nor entirely defined or defensible in each and every circumstance. The main problem, then, is the leap across this divide or the passage from observable exterior evidence to inferences about the interior. Consequently, "what one should be wary of," Derrida (2008, 79) writes by way of a reading of Descartes's *Discourse on Method*, "is the passage from outside to inside, belief in the possibility of inducing from this *exterior* resemblance an *interior* analogy, namely, the presence in the animal of a soul, of sentiments and passions like our own." Unlike Singer, who appears to tolerate the less-than-scientific approaches of physiognomy or expression, Descartes, on this account at least, "shows himself to be very prudent" (Derrida 2008, 79) by refusing to admit anything that requires conjecture, inference, or a leap of faith.

Although this "passage from the outside to the inside" (ibid.) runs into significant epistemological difficulties, this does not necessarily discount or foreclose efforts to consider seriously the moral standing of nonhuman animals. As Donna Haraway (2008, 226) argues, "the philosophic and literary conceit that all we have is representations and no access to what animals think and feel is wrong. Human beings do, or can, know more than we used to know, and the right to gauge that knowledge is rooted in historical, flawed, generative cross-species practices." Haraway affirms that the standard philosophical problem, "climbing into heads, one's own or others', to get the full story from the inside" (ibid.), is in principle not possible. But this "other minds problem" does not, she contends, foreclose efforts to understand others or excuse our responsibilities to them. In making this statement, Haraway directly confronts and contests the epistemological restraint that had been exercised since at least the time of Descartes, the thinker who Derrida had singled out for his methodological "prudence." In fact, it is on this point that Haraway's *When Species Meet* encounters and contests Derrida's *The Animal That Therefore I Am.*

Perhaps the most striking and visible point of contact and contrast between these two efforts can be found in their choice of exemplary animal. Whereas Haraway is principally concerned with dogs, Derrida has cats. Or more precisely stated, a cat—a small, female cat who on one particular occasion confronts him in the bathroom (Derrida 2008, 5). Interestingly, to say this in the Polish language—*On ma kota*—translates literally as "he has cats," but it also functions as an idiomatic expression commonly used to indicate mental derangement and instability. (And the thinking behind this particular idiom makes some intuitive sense insofar as anyone who has a number of cats in the house must be a bit "off.") According to Haraway, Derrida is not necessarily crazy; he simply does not go far enough in the examination of his encounter with this particular animal. Although the philosopher, Haraway (2008, 19–20) contends, "understood that actual animals look back at actual human beings" and that the "key question" is not "whether the cat could 'speak' but whether it is possible to know what *respond* means and how to distinguish a response from a reaction," he did not take this meeting with his cat far enough. "He came," Haraway writes, "right to the edge of respect, of the move to *respecere*, but he was side tracked by his textual canon of Western philosophy and literature" (ibid., 20).

According to Haraway's reading, it is because the philosopher got distracted, in fact has always and already been distracted, by words, and written words at that, that "Derrida failed a simple obligation of companion species; he did not become curious about what the cat might actually be doing, feeling, thinking, or perhaps making available to him in looking back at him that morning" (ibid.). Derrida, therefore, unfortunately left "unexamined the practices of communication outside the writing technologies he did know how to talk about" (ibid., 21). This critique of Derridian philosophy has a certain seductive quality to it, mainly because it mobilizes one of the popular and persistent criticisms of Derrida's entire enterprise, namely, his seemingly stubborn insistence (articulated again and again, in text after text) that "there is nothing outside the text" (Derrida 1988, 148). In effect, Haraway argues that Derrida, in this crucial and important work on the question of the animal, did what he always does. He got himself tangled up in the textual material of the Western philosophical canon, specifically the writings of Descartes, Levinas, Heidegger, and Lacan, and therefore missed a unique opportunity to engage with this cat—a real individual cat that had confronted him at a particular time and in a particular place outside the text. "I am," Haraway (2008, 23) concludes speculating about the private, interior life of Derrida the man, "prepared to believe that he did know how to greet this cat and began each morning in that mutually responsive and polite dance, but if so, that embodied mindful encounter did not motivate his philosophy in public. That is a pity."

What Haraway (2008, 26) proposes in response to this "pitiful failure" and fundamental lack of respect is an alternative notion of "communication," which she, following Gregory Batson, calls "non-linguistic embodied communication." Haraway, however, is careful to avoid the metaphysical trappings and pitfalls that are typically associated with this concept. For her, "non-linguistic embodied communication" is nothing like Jean-Jacques Rousseau's (1966, 6) "language of gesture," which, as Derrida had pointed out in *Of Grammatology*, remains firmly situated in and supportive of logocentrism; physiognomy's "language of the body, the expression of the subject's interior in his spontaneous gestures" (Žižek 2008b, 235); or the concept of nonverbal communication as it has developed in the discipline of communication studies. On the contrary, Haraway furnishes a formulation that, borrowing from the innovations of Emmanuel Levinas, is oriented otherwise. "The truth or honesty of nonlinguistic embodied communication depends on looking back and greeting the significant others, again and again. This sort of truth or honesty is not some trope-free, fantastic kind of natural authenticity that only animals can have while humans are defined by the happy fault of lying denotatively and knowing it. Rather, this truth telling is about co-constitutive naturalcultural dancing, holding in esteem, and regard open to those who look back reciprocally" (Haraway 2008, 27).

For Haraway, then, "non-linguistic embodied communication" is not some romantic notion of a direct mode of immediate concourse through bodily expression. It is neither trope-free nor a fantastic kind of "natural authenticity." It is instead a reciprocal exchange situated in the meeting of the gaze of an other. It is a "co-constitutive naturalcultural dancing" illustrated by, as Haraway presents it in considerable detail, the demanding sport of canine agility. And the operative question in these circumstances is not Bentham's "Can they suffer?" but "Can animals play? Or work? And even, can I learn to play with *this* cat?" (Haraway 2008, 22). In these playful encounters, Haraway emphasizes, the participants "do not precede the meeting" (ibid., 4) but first become who and what they are in the course of their interactions with each other. This reconceptualization of communication, where the interacting subjects are a product of the relationship and not some preexisting substance, clearly has promise for both sides of the "companion species" relationship, and Haraway describes it in a way that is careful to avoid simply slipping back into the language of metaphysics and the metaphysics of language.

Despite this promising development, however, Haraway's account redeploys that other metaphysical privilege—the privileging of vision, the eyes, and the gaze of the other. It is only those others who look back with eyes that are capable of meeting her eyes "face-to-face in the contact zone" (ibid., 227) that are considered to be capable of engaging in this kind of nonlinguistic communication. For Haraway, then, companion species are, in more ways than one, indissolubly connected to optics:

In recent speaking and writing on companion species I have tried to live inside the many tones of regard/respect/seeing each other/looking back at/meeting/optic-haptic encounter. Species and respect are in optic/haptic/affective/cognitive touch: they are at table together; they are messmates, companions, in company, *cum panis*. I also love the oxymoron inherent in "species"—always both logical type and relentless particular, always tied to *specere* and yearning/looking toward *respecere*. . . . The ethical regard that I am trying to speak and write can be experienced across many sorts of species differences. The lovely part is that we can know only by looking and by looking back. *Respecere.* (Ibid., 164)

This formulation, whether intended or not, has the effect of privileging particular kinds of animals as companion species, dogs for instance, but even some mice and cats, where the eyes are situated on the face in such a way as to be able to meet our gaze, and tends to exclude anything that does not and is structurally unable to come eye to eye or face to face with the human subject. The "ethical regard" that occupies Haraway, therefore, is something that is exclusively situated in the eyes, the proverbial window to the soul. It is about looking and looking back at each other that ultimately matters. Consequently, Haraway's ethics of respect for companion species not only capitalizes on the basic innovations of Levinasian ethics, which characterizes moral consideration as the face-to-face encounter with the Other, but also inherits one of its persistent and systemic difficulties—a conceptualization of "face" that remains, if not human, then at least humanist. Although the Other who occupies the pages of *When Species Meet* is no longer exclusively human, he/she/it is still characterized in terms that make exclusive decisions about who or what will count as other. In response to Haraway's critique, then, it might be said that Derrida does not necessarily come up short in his analysis but deliberately hesitates, in response to the intervention of a particular cat, to reproduce the exclusive decisions and operations that have character-ized anthropocentric metaphysics. Consequently, it may be the case that Derrida is in fact more respectful of the animal other and other kinds of animals than Haraway gives him credit for.

2.3.3   Ethical Problems

Beginning with Taylor's deliberately sarcastic *Vindication of the Rights of Brutes*, animal ethics has been organized and developed under the concep-tual banner of what Singer calls a "liberation movement." "A liberation movement demands an expansion of our moral horizons and an extension or reinterpretation of the basic moral principle of equality. Practices that were previously regarded as natural and inevitable come to be seen as the result of an unjustifiable prejudice" (Singer 1989, 148). Expanding the boundary of existing moral horizons in order to accommodate and include previously excluded groups sounds good and appears to be beyond ques-tion. According to Calarco (2008, 127), this "'logic of liberation' . . . is such a common way of thinking about animal ethics and other progressive political movements that very few theorists or activists would bother to question its underlying premises." This approach, however, is not without its own problems and therefore cannot be insulated from critical examina-tion. One of the first critical reconsiderations is in fact presented in Taylor's *Vindication*, where the extension of moral boundaries to previously excluded groups is pursued to what Taylor had envisioned as being an absurd and unlikely conclusion. Although Taylor's reductio ad absurdum was ulti-mately directed at undermining efforts to expand rights for women, his general skepticism about "moral expansion" is not necessarily inaccurate or misguided. In fact, Calarco (2008, 128) proposes that such endeavors, as they have been deployed and developed in animal ethics, may in fact be "a mistake, perhaps the most serious mistake that has occurred in the field."

First, efforts to expand existing moral and legal frameworks to include previously excluded subjects risks logically consistency. According to Thomas Birch:

> The nub of the problem with granting or extending rights to others, a problem which becomes pronounced when nature is the intended beneficiary, is that it presupposes the existence and the maintenance of a position of power from which to do the granting. Granting rights to nature requires bringing nature into our human system of legal and moral rights, and this is still a (homocentric) system of hierarchy and domination. The liberal mission is to open participation in the system to more and more others of more and more sorts. They are to be enabled and permitted to join the ranks and enjoy the benefits of power; they are to be absorbed. But obviously a system of domination cannot grant full equality to *all* the dominated without self-destructing. (Birch 1995, 39)

The extension of existing moral rights to previously excluded groups does not in any way challenge the basic power structure of anthropocentric (or what Birch calls, using the Latin prefix instead of the Greek, *homocentric*) ethics.[2] It employs that structure and redistributes its strategies in order to incorporate and absorb previously excluded others into its organization. Doing so not only leaves the existing hierarchies and structures of domination intact but, if taken to its logical conclusion, would eventually fall apart or self-destruct. Consequently, "there is," as Calarco (2008, 128) concludes, "a peculiar irony at work when animal rights theorists and animal liberationists employ classical humanist and anthropocentric criteria to argue for granting animals certain rights of protecting them from suffering, *for it is these very criteria that have served historically to justify violence toward animals.*"

Second, as Haraway's text demonstrates, in both word and deed, animal ethics, as it has developed and is practiced, remains an exclusive undertaking. Despite the fact that, as Singer (1975, 1) had suggested, "all animals are equal," some animals have been and continue to be more equal than others. And this exclusivity is perhaps best exemplified by the work of Tom Regan. According to Regan, "the case for animal rights" does not include all animals but is limited to those species with sufficient complexity to have at least a minimal level of mental abilities similar to a human being: "The greater the anatomical and physiological similarity between given animals and paradigmatic conscious beings (i.e. normal, developed human beings), the stronger our reasons are for viewing these animals as being like us in having the material basis for consciousness; the less like us a given animal is in this respects, the less reason we have for viewing them as having a mental life" (Regan 1983, 76).

This has the effect of instituting a highly selective, potentially inconsistent, and unfortunately capricious form of ethics, where those animals judged to be closest to us—based on perceived similarities of anatomy and physiology—are included, while others are left out of consideration altogether. For this reason, the word "animal" in Regan's *The Case for Animal Rights* is limited to "mentally normal mammals of a year or more" (ibid., 78) and excludes everything else. "Although Regan," as Calarco (2008, 130) correctly points out, "has no desire to use his theory to create a new set of exclusions that will place those animals not having these traits outside the scope of moral concern (he argues instead for a charitable approach to line drawing), this is precisely its effect." Consequently, Singer does not know to what extent he was correct. He does not know with what precision he had identified the fundamental problem with his own brand of patient-oriented ethics, when he wrote the following: "One should always be wary of talking of 'the last remaining form of discrimination.' If we have learnt anything from the liberation movements, we should have learnt how difficult it is to be aware of latent prejudice in our attitudes to particular groups until this prejudice is forcefully pointed out" (Singer 1989, 148). Animal ethics, for all its promising innovations, remains an exclusive undertaking that has its own set of latent prejudices.

Finally, and perhaps most importantly, developments in animal ethics and animal rights philosophy, although opening up the possibility of including at least some animals within the moral community, continue to exclude the machine. If, as Regan (1999, xii) had argued, the animal had been traditionally excluded from the canonical works of moral philosophy, then it is the machine that is marginalized by and excluded from the recent efforts of animal rights philosophy. In the process of deciding "where to draw the line between those animals that are, and those that are not, conscious or aware," Regan (1983, 76) inevitably relies on the figure of the machine as the paradigmatic case of the excluded other. "Because some animals frequently differ from us in quite fundamental ways in these respects, it is not unreasonable to view them as utterly lacking in consciousness. Like automatic garage doors that open when they register an electronic signal, or like the pinball machine that registers the overly aggressive play of a competitor and lights up 'Tilt!' some animals may be reasonably viewed as making their 'behavioral moves' in the world without any awareness of it" (ibid.). Despite Regan's stanch anti-Cartesianism, his work remains indebted to and informed by the figure of the animal-machine. Specifically those nonmammalian animals that operate more like an automatic mechanism than a truly sentient creature are, in Regan's estimation, justifiably excluded from moral consideration, because they simply react following preprogrammed instructions and give no indication of being aware of anything.

Regan's dividing line, therefore, differs little from the Cartesian tradition that he sought so vehemently to contest. Whereas Descartes divided human beings (even the most mentally deficient of human beings) from the animal-machine, Regan divides sentient mammals, which it is important to remember include some but not all human beings (e.g., the "profoundly mentally retarded," "mentally impoverished," and "babies less than one year old"), from those other animals that remain mere organic/biological mechanisms. What is interesting about this decision is not only that Regan continues to justify the exclusion of some animals by equating them with machines but the fact that the machine is without any question or critical hesitation situated outside the space of moral consideration tout court. When moral exclusions are enacted or when the line comes to be drawn, it is the machine that always and already occupies the position of the excluded other. In other words, the machine is not just one kind of excluded other; it is the very mechanism of the exclusion of the other.

This unquestioned exclusivity is not something that is limited to Regan's particular approach to animal ethics, but can also be found in the literature of AI and robotics and in recent critical assessments of animal rights philosophy. The former finds articulation in what Steve Torrance (2008, 502) calls the "organic view of ethical status." Although not necessarily supporting the position, Torrance argues that the organic view, which appears in a number of different versions and forms, needs to be taken seriously in the future development of the field of machine ethics. As Torrance characterizes it, the organic view includes the following five related components:

a) There is a crucial dichotomy between beings that possess organic or biological characteristics, on the one hand, and "mere" machines on the other.

b) It is appropriate to consider only a genuine organism (whether human or animal; whether naturally occurring or artificially synthesized) as being a candidate for intrinsic moral status—so that nothing that is clearly on the machine side of the machine-organism divide can coherently be considered as having any intrinsic moral status.

c) Moral thinking, feeling and action arises organically out of the biological history of the human species and perhaps many more primitive species which may have certain forms of moral status, at least in prototypical or embryonic form.

d) Only beings, which are capable of sentient feeling or phenomenal awareness could be genuine subjects of either moral concern or moral appraisal.

e) Only biological organisms have the ability to be genuinely sentient or conscious. (Torrance 2008, 502–503)

In this way, Torrance, although not directly engaged in the debates and discussions concerning animal rights philosophy, provides an articulation of moral considerability that is virtually identical to what has been advanced in the field of animal ethics. Like Regan's decision concerning animal rights, the organic view, at least as it is characterized in Torrance's article, draws a line of demarcation, instituting a dichotomy that distinguishes one category of entities from another. On the one side, there are organic or biological organisms, either naturally occurring or synthetically developed, that are sentient and therefore legitimate subjects of moral consideration. On the other side, there are mere machines—mechanisms that have no moral standing whatsoever. Consequently, as Torrance explicitly recognizes, this way of dividing things up would "definitely exclude robots from having full moral status" (ibid., 503). And it is precisely by mobilizing this perspective, although it is not always identified with the generic term "the organic view," that researchers, scientists, and engineers have typically explained and justified the exclusion of machines from serious moral consideration. Although the details might differ significantly, the basic argument remains remarkably consistent: machines cannot be legitimate moral subjects, because they are not alive.

The machine is also marginalized, as a kind of collateral damage, in recent efforts to reassess and critique the exclusive strategies that have characterized animal rights philosophy. In these cases, what is important is not so much what is explicitly indicated about the machine but a conspicuous absence that is often marked quite literally by a lack of consideration. Matthew Calarco's *Zoographies* (2008, 3), for example, has a great deal to say about the "human–animal distinction," but it remains virtually silent when it comes to other forms of otherness, namely, that of the machine. This silence is evident, to employ distinctly Derridian (1982, 65) language, in the trace of an erasure. That is, the exclusion of the machine from consideration within the text becomes manifest in the form of a trace that is left by its having been crossed out or removed from the text. Calarco, for instance, concludes his investigation of "the question of the animal" by quoting one of the more famous statements from Donna Haraway's influential "A Cyborg Manifesto": "By the late twentieth century . . . the boundary between human and animal is thoroughly breached. The last beachheads of uniqueness have been polluted if not turned into amusement parks—language, tool use, social behavior, mental events, nothing really convincingly settles the separation of human and animal. And many people no longer feel the need for such a separation" (Calarco 2008, 148). Calarco draws on and employs this passage in an effort, as he describes it, to "resolutely refuse the comfort and familiarity of the human–animal distinction" (ibid.)—a distinction that he finds stubbornly persistent and indelible even in the writings of an innovative critical thinker like Derrida. What is interesting in this particular citation of Haraway's text, however, is what Calarco decides to exclude and leave out.

For Haraway, at least in the pages of "A Cyborg Manifesto," the boundary breakdown between the human and the animal is immediately succeeded by and related to "a second leaky distinction," namely, that situated between "animal-human (organism) and machine": "Late twentieth-century machines have made thoroughly ambiguous the difference between natural and artificial, mind and body, self-developing and externally designed, and many other distinctions that used to apply to organisms and machines. Our machines are disturbingly lively, and we ourselves frighteningly inert" (Haraway 1991, 152). The "Manifesto," therefore, addresses itself to a complex and multifaceted boundary breakdown that involves and contaminates all aspects of the human–animal–machine distinction. Calarco, however, restricts his critical analysis to an investigation of the human–animal distinction and, in the process, effectively excludes the machine from consideration. And this exclusive decision becomes evident in the way he cuts off the quotation of Haraway's text. In deciding to make the incision where he did, Calarco quite literally cuts the machine out.

But Haraway, at least in her recent publications, does not do much better. Despite an emphasis in the "Manifesto" on conceptual pollutions and the blurring of the boundary that had customarily distinguished organisms from machines, her latest work, save a brief consideration of the comic potential contained in the nominal coincidence of the words "lapdog" and "laptop," appears to be more interested in redrawing a distinction between those "critters" (her word) who occupy the contact zone where species meet—"actual animals and people looking back at each other" (Haraway 2008, 42) with respect in a face-to-face encounter—and "machines whose *reactions* are of interest but who have no *presence*, no face, that demands recognition, caring, and shared pain" (ibid., 71). Despite all the promises that appear to be advanced by these recent ruminations on and innovations in moral thinking, the exclusion of the machine appears to be the last socially accepted moral prejudice.

For these reasons, animal ethics, in whatever form it is articulated and developed, is an exclusive undertaking, one that operationalizes and enacts prejudicial decisions that are just as problematic as those anthropocentric theories and practices that it had contested and hoped to replace. This conclusion, however, may not be entirely accurate or attentive to the nuances of the project of animal rights philosophy. In fact, it proceeds from and is possible only on the basis of two related assumptions. On the one hand, it could be argued that animal rights philosophy does not necessarily have any pretensions to be all inclusive. Despite the fact that Taylor (1966, 10) advanced the idea of "the equality of all things, with respect to their intrinsic and real dignity and worth" and Calarco (2008, 55) makes a strong case for "a notion of *universal ethical consideration*, that is, an agnostic form of ethical consideration that has no a priori constraints or boundaries," mainstream animal rights philosophy, at least as represented by Singer, Regan, and others, makes no commitment to this kind of totalizing universality. Unlike environmental ethics, which has, especially through the work of Birch (1993), sought to formulate an ethics of "universal consideration," animal ethics never conceived of itself as an ethics of everything. Derrida, in fact, cautions against uncritical use of the universal, all-encompassing term "Animal":

A critical uneasiness will persist, in fact, a bone of contention will be incessantly repeated throughout everything that I wish to develop. It would be aimed in the first place, once again, at the usage, in the singular, of a notion as general as "The

Animal," as if all nonhuman living things could be grouped within the common sense of this "commonplace," the Animal, whatever the abyssal differences and structural limits that separate, in the very essence of their being, all "animals," a name that we would therefore be advised, to begin with, to keep within quotation marks. (Derrida 2008, 34)

Animal rights philosophy, therefore, neither is nor aims to provide the kind of "universal consideration" that could subsequently be faulted for having made strategic decisions about who or what comes to be included and/or excluded from the moral community. Although animal rights philosophy began and remains critical of the exclusionary gestures of traditional forms of anthropocentric ethics, it does not follow from this that it must be an all-inclusive effort that does not or may not make additional, exclusive decisions.

On the other hand, the exclusion of other forms of otherness, like the machine, is only a problem if and to the extent that animals and machines share a common, or at least substantially similar, ontological status and remain effectively indistinguishable. This is precisely the argument advanced by Descartes's anthropocentric metaphysics, which draws a line of demarcation between the human subject, the sole creature capable of rational thought, and its nonhuman others, both animals and machines. In fact, for Descartes, animals and machines are, on this account, essentially interchangeable, and this conclusion is marked, quite literally within the space of the Cartesian text, by the (in)famous hyphenated compound *animal-machine*. Considered from a perspective that is informed and influenced by this Cartesian figure, animal rights philosophy might appear to be incomplete and insufficient. That is, efforts to extend moral consideration to nonhuman animals unfortunately do not consider the other side of the animal other—the machine. Or as I have argued elsewhere, "Even though the fate of the machine, from Descartes on, was intimately coupled with that of the animal, only one of the pair has qualified for ethical consideration. This exclusion is not just curious; it is illogical and indefensible" (Gunkel 2007, 126).

This conclusion, however, is only possible if one assumes and buys the association of the animal and machine, formulated in terms of either the Cartesian animal-machine or the somewhat weaker affiliation that Levy (2009, 213) marks with the term "robot-animal analogy," which animal rights philosophy does not. In fact, philosophers working on the animal question, from Singer and Regan to Derrida and Calarco, remain critical of, if not vehemently oppose to, the Cartesian legacy. In effect, their efforts target the conjoining hyphen in the animal-machine and endeavor to draw new lines of distinction that differentiate the one from the other. And the deciding factor is, almost without exception, suffering. According to Singer, for example, to remain within the Cartesian framework requires that one risk denying the very real and empirically demonstrated fact that animals can and do experience pain: "Although the view that animals are automata was proposed by the seventeenth-century French philosopher René Descartes, to most people, then and now, it is obvious that if, for example, we stick a sharp knife into the stomach of an unanaesthetized dog, the dog will feel pain" (Singer 1975, 10).

Regan follows suit, arguing that Descartes, as a consequence of his philosophical position, must have denied the reality of animal suffering. "Despite appearances to the contrary," Regan (1983, 3) writes, "they [animals] are not aware of anything, neither sights nor sounds, smells nor tastes, heat nor cold; they experience neither hunger nor thirst, fear nor rage, pleasure nor pain. Animals are, he observes at one point, like clocks: they are able to do some things better than we can, just as a clock can keep better time; but, like the clock, animals are not conscious." Although Cartesian apologists, like John Cottingham (1978) and Peter Harrison (1992), have argued that this characterization of Descartes is something of a caricature and not entirely accurate or justified, the fact of the matter is that animals and machines, within the field of animal rights philosophy at least, have been successfully distinguished in terms of sentience, specifically the feeling of pain. Whereas animals, like human beings, appear to be able to experience pain and pleasure, mechanisms like thermostats, robots, and computers, no matter how sophisticated and complex their designs, effectively feel nothing. Although it is possible to draw some rather persuasive analogical connections between animals and machines, "there is," as David Levy (2009, 214) concludes, "an extremely important difference. Animals can suffer and feel pain in ways that robots cannot."

2.3.4 Methodological Problems

If the modus operandi of animal ethics is something derived from and structured according to Bentham's question "Can they suffer?" it seems that the exclusion of the machine is entirely reasonable and justified. And this will be true as long as there is no mechanism that is able to or even appears to experience pain or some other sensation. But what if the situation were otherwise? As Derrida (2008, 81) recognizes, "Descartes already spoke, as if by chance, of a machine that simulates the living animal so well that it 'cries out that you are hurting it.'" This comment, which appears in a brief parenthetical aside in the *Discourse on Method*, had been deployed in the course of an argument that sought to differentiate human beings from the animal by associating the latter with mere mechanisms— what Derrida (2008, 79) calls the "hypothesis of the automatons." But the comment can, in light of the procedures and protocols of animal ethics, be read otherwise. That is, if it were indeed possible to construct a machine that did exactly what Descartes had postulated, that is, "cry out that you are hurting it," would we not also be obligated to conclude that such a mechanism was sentient and capable of experiencing pain? This is, it is important to note, not just a theoretical point or speculative thought experiment. Robotics engineers have, in fact, not only constructed mechanisms that synthesize believable emotional responses (Bates 1994; Blumberg, Todd, and Maes 1996; Breazeal and Brooks 2004), like the dental-training robot Simroid "who" cries out in pain when students "hurt" it (Kokoro 2009), but also systems capable of "experiencing" something like pleasure and pain.

The basic design principle behind this approach was already anticipated and explained in Čapek's influential *R.U.R.*, the 1920 stage-play that fabricated and first introduced the term "robot":

Dr. Gall: That's right. Robots have virtually no sense of physical pain, as young Rossum simplified the nervous system a bit too much. It turns out to have been a mistake and so we're working on pain now.

Helena: Why . . . Why . . . if you don't give them a soul why do you want to give them pain?

Dr. Gall: For good industrial reasons, Miss Glory. The robots sometimes cause themselves damage because it causes them no pain; they do things such as pushing their hand into a machine, cutting off a finger or even smashing their heads in. It just doesn't matter to them. But if they have pain it'll be an automatic protection against injuries.

Helena: Will they be any the happier when they can feel pain?

Dr. Gall: Quite the opposite, but it will be a technical improvement. (Čapek 2008, 28–29)

The efforts of Čapek's Dr. Gall are not, however, limited to the pages of science fiction. They have increasingly become science fact and an important aspect in robotics research and engineering. Hans Moravec (1988, 45), for instance, has made a case for "pleasure" and "pain" as adaptive control mechanisms for autonomous robotic systems. Since it is difficult, if not impossible, to program a robot to respond to all circumstances and eventualities, it is more effective to design systems that incorporate some kind of "conditioning mechanism." "The conditioning software I have in mind," Moravec writes, "would receive two kinds of messages from anywhere within the robot, one telling of success, the other of trouble. Some— for instance indications of full batteries, or imminent collisions—would be generated by the robot's basic operating system. Others, more specific to accomplishing particular tasks, could be initiated by applications programs for those tasks. I'm going to call the success messages 'pleasure' and the danger messages 'pain.' Pain would tend to interrupt the activity in progress, while pleasure would increase its probability of continuing" (ibid.).

Although the application of the terms "pleasure" and "pain" in this circumstance could be interpreted, as Frank Hoffmann (2001, 135) argues, as a "gross abuse of ethology terminology," the fact is that AI researchers and robotics engineers have successfully modeled emotions and constructed mechanisms with the capacity to react in ways that appear to be sentient. In a paper provocatively titled "When Robots Weep," Juan D. Velásquez (1998) describes a computational model of emotions called *Cathexis* and its implementation in a virtual autonomous agent named Yuppy. Yuppy is a doglike creature that is designed to behave in ways that simulate the behavior of an actual pet dog.

Yuppy produces emotional behaviors under different circumstances. For instance, when its Curiosity drive is high, Virtual Yuppy wanders around, looking for the synthetic bone which some humans carry. When it encounters one, its level of Happiness increases and specific behaviors, such as "wag the tail" and "approach the bone" become active. On the other hand, as time passes by without finding any bone, its Distress level rises and sad behaviors, such as "droop the tail," get executed. Similarly, while wandering around, it may encounter dark places which will elicit fearful responses in which it backs up and changes direction. (Velásquez 1998, 5)

If Singer's approach, which makes inferences about internal states based on the appearance of external indicators, were consistently applied to this kind of robotic entity, one might be led to conclude that such mechanisms do in fact experience something like pleasure and pain and are, on that account, minimally sentient (at least as far as Singer defines the term). In fact, it is precisely on the basis of this kind of inference that robotics engineers and AI researchers like Velásquez have routinely applied terms like "curiosity," "happiness," and "fear" to describe artificial autonomous agents. There is, however, an important distinction that, according to Singer, significantly complicates matters and forecloses such conclusions: "We know that the nervous systems of other animals were not artificially constructed to mimic the pain behavior of humans, as a robot might be artificially constructed" (Singer 1975, 12).

This seemingly simple and apparently straightforward statement leverages two conceptual oppositions that have been in play since at least Plato—nature versus artifice and real versus imitation. Animals and humans, Singer argues, can experience real pain, because they are the product of natural selection and are not technological artifacts. Although it is possible to program a robot or other device to mimic what looks like pleasure or pain, it only imitates these sensations and does not experience real pain or pleasure as such. It is possible, Singer (1975, 11) writes in that passage which mimics virtually every element of the Cartesian automaton hypothesis, that our best friend is really just a "cleverly constructed robot" designed to exhibit the outward appearance of experiencing pain but is in fact no more sentient than any other mindless mechanism. Or as Steve Torrance (2008, 499) explains, "I would not be so likely to feel moral concern for a person who behaved as if in great distress if I came to believe that the individual had no capacity for consciously feeling distress, who was simply exhibiting the 'outward' behavioural signs of distress without the 'inner' sentient states." There are, therefore, concerted efforts to differentiate between entities that are able to simulate the outward signs of various emotional states, what Torrance calls "non-conscious behavers" (ibid.), and those entities that really do experience the inner sentient state of having an experience of pain as such. To formulate it in distinctly metaphysical terms, external appearances are not the same as the true inner reality.

Although coming at this issue from an entirely different direction, AI researchers and robotics engineers employ similar conceptual distinctions (e.g., outside–inside, appearance–real, simulation–actual). Perhaps the most famous version of this in the field of AI is John Searle's "Chinese room." This intriguing and influential thought experiment, introduced in 1980 with the essay "Minds, Brains, and Programs" and elaborated in subsequent publications, was offered as an argument against the claims of strong AI. "The argument," Searle writes in a brief restatement, "proceeds by the following thought experiment":

Imagine a native English speaker who knows no Chinese locked in a room full of boxes of Chinese symbols (a data base) together with a book of instructions for manipulating the symbols (the program). Imagine that people outside the room send in other Chinese symbols which, unknown to the person in the room, are questions in Chinese (the input). And imagine that by following the instructions in the program the man in the room is able to pass out Chinese symbols which are correct answers to the questions (the output). The program enables the person in the room to pass the Turing Test for understanding Chinese but he does not understand a word of Chinese. (Searle 1999, 115)

The point of Searle's imaginative albeit ethnocentric[3] illustration is quite simple—simulation is not the real thing. Merely shifting symbols around in a way that looks like linguistic understanding is not really an understanding of the language. A computer, as Terry Winograd (1990, 187) explains, does not really understand the linguistic tokens it processes; it merely "manipulates symbols without respect to their interpretation." Or, as Searle concludes, registering the effect of this insight on the standard test for artificial intelligence: "This shows that the Turing test fails to distinguish real mental capacities from simulations of those capacities. Simulation is not duplication" (Searle 1999, 115).

A similar point has been made in the consideration of other mental capacities, like sentience and the experience of pain. Even if, as J. Kevin O'Regan (2007, 332) writes, it were possible to design a robot that "screams and shows avoidance behavior, imitating in all respects what a human would do when in pain . . . All this would not guarantee that to the robot, there was actually *something it was like* to have the pain. The robot might simply be going through the motions of manifesting its pain: perhaps it actually feels nothing at all. Something extra might be required for the robot to *actually experience* the pain, and that extra thing is *raw feel*, or what Ned Block calls *Phenomenal Consciousness.*" For O'Regan, programmed behavior that looks a lot like pain is not really an experience of pain. And like Searle, he asserts that something more would be needed in order for these appearances of the feeling of pain to be actual pain.

These thought experiments and demonstrations, whether it is ever explicitly acknowledged as such or not, are different versions of the Socratic argument against the technology of writing that was presented at the end of Plato's *Phaedrus*. According to Socrates, a written text may offer the appearance of something that looks like intelligence, but it is not on this account actually intelligent. "Writing," Plato (1982, 275d) has Socrates say, "has this strange quality, and is very much like painting; for the creatures of painting stand like living beings, but if one asks them a question, they preserve a solemn silence. And so it is with written words; you might think they spoke as if they had intelligence, but if you question them, wishing to know about their sayings, they always say only one and the same thing." According to this Socratic explanation, a technological artifact, like a written document, often gives appearances that might lead one to conclude that it possessed something like intelligence; but it is not, on the basis of that mere appearance, actually intelligent. If interrogated, the written document never says anything new or innovative. It only says one and the same thing ad infinitum. It is, therefore, nothing more than a dead artifact that can only reproduce preprogrammed instructions, giving the appearance of something that it really does not possess.

Drawing a distinction between the mere appearance of something and the real thing as it really is in itself is a persuasive distinction that has considerable philosophical traction. This is, as any student of philosophy will immediately recognize, the basic configuration typically attributed to Platonic metaphysics. For mainstream Platonism, the real is situated outside of and beyond phenomenal reality. That is, the real things are located in the realm of supersensible ideas—$\varepsilon\iota\delta o\varsigma$ in Plato's Greek—and what is perceived by embodied and finite human beings are derived and somewhat deficient apparitions. This "doctrine of the forms," as it eventually came to be called, is evident, in various forms, throughout the Platonic corpus. It is, for example, illustrated at the center of the *Republic* with the allegory of the cave. The allegory, ostensibly an image concerning the deceptive nature of images, distinguishes between the mere shadowy apparition of things encountered in the subterranean cavern and the real things revealed as such under the full illumination of the sun. For this ontological difference, as it is commonly called, to show itself as such, however, one would need access not just to the appearance of something but to the real thing as it really is in itself. In other words, the appearance of something is only able to be recognized as such and to show itself as an appearance on the basis of some knowledge of the real thing against which it is compared and evaluated.

Although this sounds a bit abstract, it can be easily demonstrated by way of a popular television game show from the so-called golden age of television in the United States. The show, *To Tell the Truth*, was created by Bob Stewart, produced by the highly successful production team of Mark Goodson and Bill Todman (arguably the Rogers and Hammerstein of the television game show industry), and ran intermittently on several U.S. television networks since its premier in the mid-1950s. *To Tell the Truth* was a panel show, which, like its precursor *What's My Line?* (1950–1967), featured a panel of four celebrities, who were confronted with a group of three individuals or challengers.[4] Each challenger claimed to be one particular individual who had some unusual background, notable life experience, or unique occupation. The celebrity panel was charged with interrogating the trio and deciding, based on the responses to their questions, which one of the three was actually the person he or she purported to be—who, in effect, was telling the truth. In this exchange, two of the challengers engaged in deliberate deception, answering the questions of the celebrity panel by pretending to be someone they were not, while the remaining challenger told the truth. The "moment of truth" came at the game's conclusion, when the program's host asked the pivotal question, "Will the real so-and-so please stand up?" at which time one of the three challengers stood. In doing so, this one individual revealed him- or herself as the real thing and exposed the other two as mere imposters. This demonstration, however, was only possible by having the real thing eventually stand up and show him- or herself as such.

Demonstrations, like Searles's Chinese room, that seek to differentiate between the appearance of something and the real thing as it "really" is, inevitably need some kind of privileged and immediate access to the real as such and not just how it appears. In order to distinguish, for example, between the appearance of experiencing pain and the reality of an actual experience of pain, researchers would need access not just to external indicators that look like pain but to the actual experiences of pain as it occurs in the mind or body of another. This requirement, however, has at least two fundamental philosophical problems. First, this procedure, not surprisingly, runs into the other minds problem. Namely, we cannot get into the heads of other entities—whether human being, nonhuman animal, alien life form, or machine—to know with any certainty whether they actually experience whatever it is they appear to manifest to us. But the situation is actually more complicated and widespread than this particular and seemingly perennial problem from the philosophy of mind. This is because human knowledge, according to the critical work of Immanuel Kant, is absolutely unable to have access to and know anything about something as it really is in itself.

Kant, following the Platonic precedent, differentiates between an object as it appears to us (finite and embodied human beings) through the mediation of the senses and the thing as it really is in itself (*das Ding an sich*). "What we have meant to say," Kant (1965, A42/B59) writes in the opening salvo of the *Critique of Pure Reason*, "is that all our intuition is nothing but the representation of appearance; that the things which we intuit are not in themselves what we intuit them as being, nor their relations so constituted in themselves as they appear to us." This differentiation installs a fundamental and irreconcilable split whereby "the object is to be taken in *a two fold sense*, namely as appearance and as thing in itself" (ibid., Bxxvii). Human beings are restricted to the former, while the latter remains, for us at least, forever unapproachable. "What objects may be in themselves, and apart from all this receptivity of our sensibility, remains completely unknown to us. We know nothing but our mode of perceiving them—a mode, which is peculiar to us, and not necessarily shared in by every being, though, certainly by every human being" (ibid., A42/B59).

Despite the complete and absolute inaccessibility of the thing itself, Kant still "believes" in its existence: "But our further contention must also be duly borne in mind, namely that though we cannot *know* these objects as things in themselves, we must yet be in a position at least to think them as *things* in themselves; otherwise we should be landed in the absurd conclusion that there can be appearances without anything that appears" (ibid., Bxxvi). Consequently, Kant redeploys the Platonic distinction between the real thing and its mere appearances, adding the further qualification that access to the real thing is, if we are absolutely careful in defining the proper use and limits of our reason, forever restricted and beyond us. What this means for the investigation of machine patiency (and not just machine patiency but patiency in general) is both clear and considerably unsettling. We are ultimately unable to decide whether a thing—anything animate, inanimate, or otherwise—that appears to feel pain or exhibits some other kind of inner state has or does not have such an experience in itself. We are, in other words, unable to jump the chasm that separates how something appears to us from what that thing is in itself. Although this might sound cold and insensitive, this means that if something looks like it is in pain, we are, in the final analysis, unable to decide with any certainty whether it really is in pain or not.

Second, not only is access to the thing as it is in itself difficult if not impossible to achieve, but we may not even be able to be certain that we know what "pain" is in the first place. This second point is something that is questioned and investigated by Daniel Dennett in "Why You Can't Make a Computer That Feels Pain." In this provocatively titled essay, originally published decades before the debut of even a rudimentary working proto-type of a pain-feeling mechanism, Dennett imagines trying to disprove the standard argument for human (and animal) exceptionalism "by actually writing a pain program, or designing a pain-feeling robot" (Dennett 1998, 191). At the end of what turns out to be a rather protracted and detailed consideration of the problem, Dennett concludes that we cannot, in fact, make a computer that feels pain. But the reason for drawing this conclu-sion does not derive from what one might expect, nor does it offer any kind of support for the advocates of moral exceptionalism. According to Dennett, the fact that you cannot make a computer that feels pain is not the result of some technological limitation with the mechanism or its programming. It is a product of the fact that we remain unable to decide what pain is in the first place. The best we are able to do, as Dennett's attentive consideration illustrates, is account for the various "causes and effects of pain," but "pain itself does not appear" (ibid., 218).

In this way, Dennett's essay, which is illustrated with several intricate flow chart diagrams, confirms something that Leibniz had asserted con-cerning perceptions of any kind: "If we imagine that there is a machine whose structure makes it think, sense, and have perceptions, we could conceive of it enlarged, keeping the same proportions, so that we could enter into it, as one enters into a mill. Assuming that, when inspecting its interior, we will only find parts that push one another, and we will never find anything to explain a perception" (Leibniz 1989, 215). Like Dennett, Leibniz's thought experiment, which takes a historically appropriate mechanical form rather than one based on computational modeling, is able to identify the causal mechanisms of sensation but is not capable of locating a sensation as such. What Dennett demonstrates, therefore, is not that some workable concept of pain cannot come to be instantiated in the mechanism of a computer or a robot, either now or in the foreseeable future, but that the very concept of pain that would be instantiated is already arbitrary, inconclusive, and indeterminate. "There can," Dennett (1998, 228) writes at the end of the essay, "be no true theory of pain, and so no computer or robot could instantiate the true theory of pain, which it would have to do to feel real pain." What Dennett proves, then, is not an inability to program a computer to feel pain but our initial and persistent inability to decide and adequately articulate what constitutes the experience of pain in the first place. Although Bentham's question "Can they suffer?" may have radically reoriented the direction of moral philosophy, the fact remains that "pain" and "suffering" are just as nebulous and difficult to define and locate as the concepts they were introduced to replace.

Finally, all this talk about the possibility of engineering pain or suffering in a machine entails its own particular moral dilemma. "If (ro)bots might one day be capable of experiencing pain and other affective states," Wallach and Allen (2009, 209) write, "a question that arises is whether it will be moral to build such systems—not because of how they might harm humans, but because of the pain these artificial systems will themselves experience. In other words, can the building of a (ro)bot with a somatic architecture capable of feeling intense pain be morally justified and should it be prohibited?" If it were in fact possible to construct a machine that "feels pain" (however that would be defined and instantiated) in order to demonstrate the limits of sentience, then doing so might be ethically suspect insofar as in constructing such a mechanism we do not do everything in our power to minimize its suffering. Consequently, moral philosophers and robotics engineers find themselves in a curious and not entirely comfortable situation. One needs to be able to construct such a machine in order to demonstrate sentience and moral responsibility; but doing so would be, on that account, already to engage in an act that could potentially be considered immoral. The evidence needed to prove the possibility of moral responsibility, then, seems to require actions the consequences of which would be morally questionable at best. Or to put it another way, demonstrating the moral standing of machines might require unethical actions; the demonstration of moral patiency might itself be something that is quite painful for others.

2.4   Information Ethics

One of the criticisms of animal rights philosophy is that this moral innovation, for all its promise to intervene in the anthropocentric tradition, remains an exclusive and exclusionary practice. "If dominant forms of ethical theory," Calarco (2008, 126) concludes, "—from Kantianism to care ethics to moral rights theory—are unwilling to make a place for animals within their scope of consideration, it is clear that emerging theories of ethics that are more open and expansive with regard to animals are able to develop their positions only by making other, equally serious kinds of exclusions." Environmental and land ethics, for instance, have been critical of Singer's "animal liberation" and animal rights philosophy for including some sentient creatures in the community of moral patients while simultaneously excluding other kinds of animals, plants, and the other entities that make up the natural environment (Sagoff 1984). In response to this exclusivity, environmental ethicists have, following the precedent and protocols of previous liberation efforts like animal rights philosophy, argued for a further expansion of the moral community to include these marginalized others. "The land ethic," Aldo Leopold (1966, 239) wrote, "simply enlarges the boundaries of the community to include soils, waters, plants, and animals, or collectively, the land." Such an ethics makes a case for extending moral and even legal "rights to forests, oceans, rivers, and other so-called 'natural objects'" (Stone 1974, 9). Or as Paul W. Taylor (1986, 3) explains, "environmental ethics is concerned with the moral relations that hold between humans and the natural world. The ethical principles governing those relations determine our duties, obligations, and responsibilities with regard to the Earth's natural environment and all the animals and plants that inhabit it."

Although this effort effectively expands the community of legitimate moral patients to include those others who had been previously left out, environmental ethics has also (and not surprisingly) been criticized for instituting additional omissions. In particular, the effort has been cited for privileging "natural objects" (Stone 1974, 9) and the "natural world" (Taylor 1986, 3) to the exclusion of nonnatural artifacts, like artworks, architecture, technology, machines, and the like (Floridi 1999, 43). This exemption is evident by the fact that these other entities typically are not given any explicit consideration whatsoever. That is, they are literally absent from the material of the text, as is the case with Leopold's writing on the land ethic, which says nothing about the place of nonnatural artifacts. Or it is explicitly identified, explained, and even justified, as is the case with Taylor's *Respect for Nature*, which argues, by way of mobilizing the standard anthropological and instrumental theories, that machines do not have "a good of their own" that would need to be respected:

> The ends and purposes of machines are built into them by their human creators. It is the original purposes of humans that determine the structures and hence the teleological functions of those machines. Although they manifest goal-directed activities, the machines do not, as independent entities, have a good of their own. Their "good" is "furthered" only insofar as they are treated in such a way as to be an effective means to human ends. A living plant or animal, on the other hand, has a good of its own in the same sense that a human being has a good of its own. It is, independently of anything else in the universe, itself a center of goal-oriented activity. What is good or bad for it can be understood by reference to its own survival, health and well-being. As a living thing it seeks its own ends in a way that is not true of any teleologically structured mechanism. It is in terms of *its* goals that we can give teleological explanations of why it does what it does. We cannot do the same for machines, since any such explanation must ultimately refer to the goals their human producers had in mind when they made the machines. (Taylor 1986, 124)

What is remarkable about Taylor's explicit exclusion of the machine from his brand of environmental ethics is the recognition that such exclusivity might not necessarily apply to every kind of machine. "I should add as a parenthetical note," Taylor continues, "that this difference between mechanism and organism may no longer be maintainable with regard to those complex electronic devices now being developed under the name of artificial intelligence" (ibid., 124–125). With this brief aside, therefore, Taylor both recognizes the structural limits of environmental ethics, which does not consider the machine a legitimate moral subject, and indicates the possibility that a future moral theory may need to consider these excluded others as legitimate moral patients on par with other organisms.

One scholar who has taken up this challenge is Luciano Floridi, who advances what he argues is a new "ontocentric, patient-oriented, ecological macroethics" (Floridi 2010, 83). Floridi introduces and situates this concept by revisiting what he understands as the irreducible and fundamental structure of any and all action. "Any action," Floridi (1999, 41) writes, "whether morally loaded or not, has the logical structure of a binary relation between an agent and a patient." Standard or classic forms of ethics, he argues, have been exclusively concerned with either the character of the agent, as in virtue ethics, or the actions that are performed by the agent, as in consequentialism, contractualism, and deontologism. For this reason, Floridi concludes, classic ethical theories have been "inevitably anthropocentric" in focus, and "take only a relative interest in the patient," or what he also refers to as the "receiver" or "victim" (ibid., 41–42). This philosophical status quo has been recently challenged by animal and environmental ethics, both of which "attempt to develop a patient-oriented ethics in which the 'patient' may be not only a human being, but also any form of life" (ibid., 42).

However innovative these alterations have been, Floridi finds them to be insufficient for a truly universal and impartial ethics. "Even Bioethics and Environmental Ethics," he argues, "fail to achieve a level of complete universality and impartiality, because they are still biased against what is inanimate, lifeless, or merely possible (even Land Ethics is biased against technology and artefacts, for example). From their perspective, only what is alive deserves to be considered as a proper centre of moral claims, no matter how minimal, so a whole universe escapes their attention" (ibid., 43). For Floridi, therefore, bioethics and environmental ethics represent something of an incomplete innovation in moral philosophy. They have, on the one hand, successfully challenged the anthropocentric tradition by articulating a more universal form of ethics that not only shifts attention to the patient but also expands who or what qualifies for inclusion as a patient. At the same time, however, both remain ethically biased insofar as they substitute a biocentrism for the customary anthropocentrism. Consequently, Floridi endeavors to take the innovations introduced by bioethics and environmental ethics one step further. He retains their patient-oriented approach but "lowers the condition that needs to be satisfied, in order to qualify as a centre of moral concern, to the minimal common factor shared by any entity" (ibid.), whether animate, inanimate, or otherwise.

For Floridi this lowest common denominator is informational and, for this reason, he gives his proposal the name "Information Ethics" or IE:

IE is an ecological ethics that replaces *biocentrism* with *ontocentrism*. IE suggests that there is something even more elemental than life, namely *being*—that is, the existence and flourishing of all entities and their global environment—and something more fundamental than suffering, namely *entropy*. Entropy is most emphatically not the physicists' concept of thermodynamic entropy. Entropy here refers to any kind of *destruction* or *corruption* of informational objects, that is, any form of impoverishment of *being* including *nothingness*, to phrase it more metaphysically. (Floridi 2008, 47)

Following the innovations of bio- and environmental ethics, Floridi expands the scope of moral philosophy by altering its focus and lowering the threshold for inclusion, or, to use Floridi's terminology, the level of abstraction (LoA). What makes someone or something a moral patient, deserving of some level of ethical consideration (no matter how minimal), is that it exists as a coherent body of information. Consequently, something can be said to be good, from an IE perspective, insofar as it respects and facilitates the informational welfare of a being and bad insofar as it causes diminishment, leading to an increase in information entropy. In fact, for IE, "fighting information entropy is the general moral law to be followed" (Floridi 2002, 300).

This fundamental shift in focus opens up the field of moral consideration to many other kinds of others:

From an IE perspective, the ethical discourse now comes to concern information as such, that is not just all persons, their cultivation, well-being and social interactions, not just animals, plants and their proper natural life, but also anything that exists, from paintings and books to stars and stones; anything that may or will exist, like future generations; and anything that was but is no more, like our ancestors. Unlike other non-standard ethics, IE is more impartial and universal—or one may say less ethically biased—because it brings to ultimate completion the process of enlargement of the concept of what may count as a centre of information, no matter whether physically implemented or not. (Floridi 1999, 43)

Although they are on opposite ends of the philosophical spectrum, Floridi's *ontocentric* IE looks substantially similar to what John Llewelyn (2010, 110) proposes under the banner of "ecoethics" which is "a truly democratic ecological ethicality" that is devised by engaging with and leveraging the innovations of Emmanuel Levinas.[5] For Llewelyn (2010, 108), what really matters is *existence* as such: "Existence as such is our topic. We are treating not only of existence now as against existence in the past or in the future. But we are treating of existence in the field of ecoethical decision, that is to say where what we do can make a difference to the existence of something, where what we do can contribute to bringing about its non-existence." According to Llewlyn's argument, existence is ethically relevant. We have a moral responsibility to take the existence of others, whether currently present before us or not, into consideration in all we do or do not do. Consequently, what is considered morally "good" is whatever respects the existence of "the other being, human or non-human" (Llewelyn 2010, 109).

Conversely, what is morally "bad" is whatever contributes to its non-existence. Llewelyn, however and quite understandably given his point of departure, does not identify this by way of Floridi's reformulated version of the term "entropy." Instead, he employs a modified version of a concept derived from animal rights philosophy—*suffering*. "Suffering is not necessarily the suffering of pain. Something suffers when it is deprived of a good. But among a thing's goods is its existence. Independently of the thing's nature, of the predicates, essential or otherwise, under which it falls, is its existence. The thing's existence as such is one of the thing's goods, what it would ask us to safeguard if it could speak, and if it cannot speak, it behooves those that can speak to speak for it" (Llewelyn 2010, 107). Although not characterized in informational terms, Llewelyn's reworking of the concept "suffering" is substantially similar, in both form and function, to what Floridi had done with and indicated by the term "entropy." According to Llewelyn's argument, therefore, we have a responsibility to safeguard and respect everything that exists and to speak for and on behalf of those entities that cannot speak up for themselves, namely, "animals, trees, and rocks" (Llewelyn 2010, 110). But in referring ecoethics to speech and the responsibility to speak for those who cannot, Llewelyn's proposal remains grounded in and circumscribed by λόγος and the one entity that has been determined to possesses λόγος as its sole defining characteristic, the human being or ζῷον λόγον ἔχον. This means that Llewelyn's "ecoethics," however promising it initially appears, is still a kind of humanism, albeit a humanism that is interpreted in terms of the "humane" (Llewelyn 2010, 95). Floridi's IE has its own issues, but it is at least formulated in a way that challenges the residue of humanism all the way down.

And this challenge is fundamental. In fact, IE comprises what one might be tempted to call "the end of ethics," assuming that we understand the word "end" in its full semantic range. According to Heidegger, *end* names not just the termination of something, the *terminus* or point at which it ceases to be or runs out, but also the completion or fulfillment of its purpose or intended project—what the ancient Greeks had called τελος. "As a completion," Heidegger (1977b, 375) writes, "an end is the gathering into the most extreme possibilities." The project of IE, on Floridi's account, would bring to completion the project of what Singer (1989, 148) called "a liberation movement." Like other non-standard ethics, IE is interested in expanding membership in the moral community so as to incorporate previously excluded non-human others. But unlike these previous efforts, it is "more impartial" and "more universal." That is, it does not institute what would be additional morally suspect exclusions and its universality is more universal—that is, properly universal—then what had been instituted by either animal rights philosophy or environmental ethics. As such, IE is determined to achieve a more adequate form of moral universalism that is, as Bernd Carsten Stahl (2008, 98) points out, a fundamental aspect "that has occupied ethicists for millennia," and in so doing would, it appears, finally put an end to the seemingly endless quibbling about who or what is or should be a legitimate moral subject.

This does not mean, it should be noted, that Floridi advocates even for a second that IE is somehow fully-formed and perfect. "IE's position," he explicitly recognizes, "like that of any other macroethics, is not devoid of problems" (Floridi 2005, 29). He does, however, express considerable optimism concerning its current and future prospects. "IE strives," Floridi (2002, 302–303) writes with an eye on the not-too-distant future, "to provide a good, unbiased platform from which to educate not only computer science and ICT students but also the citizens of an information society. The new generations will need a mature sense of ethical responsibility and stewardship of the whole environment both biological and informational, to foster responsible care of it rather than despoliation or mere exploitation."

For this reason, Floridi's IE, as many scholars working in the field of ICT and ethics have recognized,[6] constitutes a compelling and useful proposal. This is because it not only is able to incorporate a wider range of possible objects (living organisms, organizations, works of art, machines, historical entities, etc.) but also expands the scope of ethical thinking to include those others who have been, for one reason or another, typically excluded from recent innovations in moral thinking. Despite this considerable advantage, however, IE is not without its critics. Mikko Siponen (2004, 279), for instance, praises Floridi's work "for being bold and anti-conventional, aimed at challenging the fundamentals of moral thinking, including what constitutes moral agency, and how we should treat entities deserving moral respect." At the same time, however, he is not convinced that IE and its focus on information entropy provides a better articulation of moral responsibility. In fact, Siponen argues that "the theory of IE is less pragmatic than its key competitors (such as utilitarianism and the universalizability theses)" (ibid., 289) and for this reason, IE is ultimately an impractical mode of practical philosophy. Stahl (2008), who is interested in contributing to "the discussion of the merits of Floridi's information ethics," targets the theory's claim to universality, comparing it to what has been advanced in another approach, specifically the discourse ethics of Jürgen Habermas and Karl-Otto Apel. The objective of "this comparison of two pertinent ethical theories" is, as Stahl sees it, to initiate "a critical discussion of areas where IE currently has room for elaboration and development" (ibid., 97).

Taking things further, Philip Brey (2008, 110) credits Floridi with introducing what is arguably "a radical, unified macroethical foundation for computer ethics and a challenging ethical theory in its own right" that moves moral philosophy "beyond both the classical anthropocentric position that the class of moral patients includes only humans, and beyond the biocentric and ecocentric positions according to which the class of moral patients consists of living organisms or elements of the ecosystem" (ibid., 109). Despite its promising innovations, however, Brey still finds IE, at least as it has been presented and argued by Floridi, to be less than persuasive. He therefore suggests a "modification" that will, in effect, allow the theory to retain the baby while throwing out the bathwater. "I will argue," Brey writes, "that Floridi has presented no convincing arguments that everything that exists has some minimal amount of intrinsic value. I will argue, however, that his theory could be salvaged in large part if it were modified from a value-based into a respect-based theory, according to which many (but not all) inanimate things in the world deserve moral respect, not because of intrinsic value, but because of their (potential) extrinsic, instrumental or emotional value for persons" (ibid.). What is interesting about Brey's proposed fix is that it reintroduces the very anthropocentric privilege that IE had contested to begin with. "Floridi could,"

Brey writes, prescribing what he believes should be done, "argue that inanimate objects, although not possessive of intrinsic value, deserve respect because of either their extrinsic value or their (actual or potential) instrumental or emotional value for particular human beings (or animals) or for humanity as a whole" (ibid., 113). In other words, what Brey offers as a fix to a perceived problem with IE is itself the very problem IE sought to address and remediate in the first place.

These responses to the project of IE can be considered "critical" only in the colloquial sense of the word. That is, they identify apparent problems or inconsistencies with IE as it is currently articulated in order to advance corrections, adjustments, modifications, or tweaks that are intended to make the system better. There is, however, a more fundamental understanding of the practice that is rooted in the tradition of critical philosophy and that endeavors not so much to identify and repair flaws or imperfections but to analyze "the grounds of that system's possibility." Such a critique, as Barbara Johnson (1981, xv) characterizes it, "reads backwards from what seems natural, obvious, self-evident, or universal in order to show that these things have their history, their reasons for being the way they are, their effects on what follows from them, and that the starting point is not a given but a construct usually blind to itself." Taking this view of things, we can say that IE has at least two *critical* problems.

First, in shifting emphasis from an agent-oriented to a patient-oriented ethics, Floridi simply inverts the two terms of a traditional binary opposition. If classic ethical thinking has been organized, for better or worse, by an interest in the character and/or actions of the agent at the expense of the patient, IE endeavors, following the innovations modeled by environmental ethics and bioethics, to reorient things by placing emphasis on the other term. This maneuver is, quite literally, a revolutionary proposal, because it inverts or "turns over" the traditional arrangement. Inversion, however, is rarely in and of itself a satisfactory mode of intervention. As Nietzsche, Heidegger, Derrida, and other poststructuralists have pointed out, the inversion of a binary opposition actually does little or nothing to challenge the fundamental structure of the system in question. In fact, inversion preserves and maintains the traditional structure, albeit in an inverted form. The effect of this on IE is registered by Kenneth Einar Himma, who, in an assessment of Floridi's initial publications on the subject, demonstrates that a concern for the patient is nothing more than the flip side of good old agent-oriented ethics. "To say that an entity X has moral standing (i.e., is a moral patient) is, at bottom, simply to say that it is possible for a moral agent to commit a wrong against X. Thus, X has moral standing if and only if (1) some moral agent has at least one duty regarding the treatment of X and (2) that duty is owed to X" (Himma 2004, 145). According to Himma's analysis, IE's patient-oriented ethics is not that different from traditional ethics. It simply looks at the agent–patient couple from the other side and in doing so still operates on and according to the standard system.

Second, IE not only alters the orientation of ethics by shifting the perspective from agent to patient, but also enlarges its scope by reducing the minimum requirements for inclusion. "IE holds," Floridi (1999, 44) argues, "that every entity, as an expression of being, has a dignity, constituted by its mode of existence and essence, which deserves to be respected and hence place moral claims on the interacting agent and ought to contribute to the constraint and guidance of his ethical decisions and behaviour. This ontological equality principle means that any form of reality (any instance of information), simply for the fact of being what it is, enjoys an initial, overridable, equal right to exist and develop in a way which is appropriate to its nature." IE, therefore, contests and seeks to replace both the exclusive anthropocentric and biocentric theories with an "ontocentric" one, which is, by comparison, much more inclusive and universal.

In taking this approach, however, IE simply replaces one form of centrism with another. This is, as Emmanuel Levinas points out, really nothing different; it is more of the same: "Western philosophy has most often been an ontology: a reduction of the other to the same by interposition of a middle or neutral term that ensures the comprehension of being" (Levinas 1969, 43). According to Levinas's analysis, the standard operating procedure/presumption of Western philosophy has been the reduction of difference. In fact, philosophy has, at least since the time of Aristotle, usually explained and dealt with difference by finding below and behind apparent variety some common denominator that is irreducibly the same. Anthropocentric ethics, for example, posits a common humanity that underlies and substantiates the perceived differences in race, gender, ethnicity, class, and so on. Likewise, biocentric ethics assumes that there is a common value in life itself, which subtends all forms of available biological diversity. And in the ontocentric theory of IE, it is being, the very matter of ontology itself, that underlies and supports all apparent differentiation. As Himma (2004, 145) describes it, "every existing entity, whether sentient or non-sentient, living or non-living, natural or artificial, has some minimal moral worth . . . in virtue of its existence." But as Levinas argues, this desire to articulate a universal, common element, instituted either by explicit defini-tion or Floridi's method of abstraction, effectively reduces the difference of the other to what is ostensibly the same. "Perceived in this way," Levinas (1969, 43) writes, "philosophy would be engaged in reducing to the same all that is opposed to it as other." In taking an ontocentric approach, therefore, IE reduces all difference to a minimal common factor that is supposedly shared by any and all entities. As Floridi (2002, 294) explains it, "the moral value of an entity is based on its ontology." Although this approach provides for a more inclusive kind of "centrism," it still utilizes a centrist approach and, as such, necessarily includes others by reducing their differences to some preselected common element or level of abstraction.

A similar criticism is advanced by environmental ethicist Thomas Birch, who finds any and all efforts to articulate some criteria for "universal con-sideration" to be based on a fundamentally flawed assumption. According to Birch, these endeavors always proceed by way of articulating some nec-essary and sufficient conditions, or qualifying characteristics, that must be met by an entity in order to be included in the community of legitimate moral subjects. And these criteria have been specified in either anthropo-logical, biological, or, as is the case with IE, ontological terms. In the tra-ditional forms of anthropocentric ethics, for example, it was the *anthropos* and the way it had been characterized (which it should be noted was always and already open to considerable social negotiation and redefini-tion) that provided the criteria for deciding who would be included in the moral community and who or what would not. The problem, Birch con-tends, is not with the particular kind of centrism that is employed or the criteria that are used to define and characterize it. The problem is with the entire strategy and approach. "The institution of *any* practice of *any* crite-rion of moral considerability," Birch (1993, 317) writes, "is an act of power over, and ultimately an act of violence toward, those others who turn out to fail the test of the criterion and are therefore not permitted to enjoy the membership benefits of the club of *consideranda*. They become fit objects of exploitation, oppression, enslavement, and finally extermination. As a result, the very question of moral considerability is ethically problematic itself, because it lends support to the monstrous Western project of planetary domination."

Considered from this perspective, IE is not as radical or innovative as it first appears. Although it contests the apparent advancements of biocentrism, which had previously contested the limits of anthropocentrism, it does so by simply doing more of the same. That is, it critiques and repairs the problems inherent in previous forms of macroethics by introducing one more centrism—ontocentrism—and one more, supposedly "final criterion" (Birch 1993, 321). In doing so, however, IE follows the same procedures, makes the same kind of decisions, and deploys the same type of gestures. That is, it contests one form of centrism by way of instituting and establishing another, but this substitution does not, in any fundamental way, challenge or change the rules of the game. If history is any guide, a new centrism and criterion, no matter how promising it might initially appear, is still, as both Birch and Levinas argue, an act of power and violence against others. IE, therefore, is not, despite claims to the contrary, sufficient for a truly radical reformulation of macroethics. It simply repackages the same old thing—putting old wine in a new bottle.

2.5 Summary

Moral patiency looks at the machine question from the other side. It is, therefore, concerned not with determining the moral character of the agent or weighing the ethical significance of his/her/its actions but with the victim, recipient, or receiver of such action. This approach is, as Hajdin (1994), Floridi (1999), and others have recognized, a significant alteration in procedure and a "nonstandard" way to approach the question of moral rights and responsibilities. It is quite literally a *revolutionary* alternative insofar as it turns things around and considers the ethical relationship not from the perspective of the active agent but from the position and viewpoint of the recipient or patient. The model for this kind of patient-oriented ethics can be found in animal rights philosophy. Whereas agent-oriented approaches have been concerned with determining whether someone is or is not a legitimate moral person with rights and responsibilities, animal rights philosophy begins with an entirely different question—"Can they suffer?" This seemingly simple and direct inquiry introduces a paradigm shift in the basic structure and procedures of moral philosophy.

On the one hand, animal rights philosophy challenges the anthropocentric tradition in ethics by critically questioning the often unexamined privilege human beings have granted themselves. In effect, it institutes something like a Copernican revolution in moral philosophy. Just as Copernicus challenged the geocentric model of the cosmos and in the process undermined many of the presumptions of human exceptionalism, animal rights philosophy challenges the established system of ethics, deposing the anthropocentric privilege that has traditionally organized the moral universe. On the other hand, the effect of this significant shift in focus means that the once-closed field of ethics is opened up to including other kinds of nonhuman others. In other words, who counts as morally significant are not just other "men" but all kinds of entities that had previously been marginalized and situated outside the gates of the moral community.

Despite this important innovation, animal rights philosophy has a less than laudable resume, and it runs into a number of significant and seemingly inescapable difficulties. First, there is a problem with terminology, one that is not merely matter of semantics but affects the underlying conceptual apparatus. Although Bentham's question effectively shifts the focus of moral consideration from an interest in determining the "person-making qualities," like (self-) consciousness and rationality, to a concern with and for the suffering of others, it turns out that "suffering" is just as ambiguous and indeterminate. Like consciousness, suffering is also one of those occult properties that admit of a wide variety of competing characterizations. To make matters worse, the concept, at least in the hands of Singer and Regan, is understood to be coextensive with "sentience," and has the effect of turning Bentham's question concerning an essential vulnerability into a new kind of mental power and capacity. In this way, sentience looks suspiciously like consciousness just formulated at what Floridi and Sanders (2003) call "a lower level of abstraction."

Second, and following from this, there is the seemingly irresolvable epistemological problem of other minds. Even if it were possible to decide on a definition of suffering and to articulate its necessary and sufficient conditions, there remains the problem of knowing whether someone or something that appears to be suffering is in fact actually doing so, or whether it is simply reacting to adverse stimulus in a preprogrammed or automatic way, or even dissimulating effects and symptoms that look like pain. Attempts to resolve these issues often conduct the debate concerning animal rights into quasi-empirical efforts or pseudo-sciences like physiognomy, where one tries to discern internal states and experiences from physiological evidence and other forms of observable phenomena. Like Descartes, animal rights philosophers unfortunately find themselves in the uncomfortable situation of being unable to decide in any credible way whether an other—whether another person, animal, or thing—actually suffers and experiences what is assumed to be pain. Although the question "Can they suffer?" effectively alters the criterion of decision, asking us to consider a different set of issues and requiring that we look for different kinds of evidence, the basic epistemological problem remains intact and unchallenged.

Beyond these substantive problems, animal rights philosophy also has significant ethical consequences. Although it effectively challenges the prejudice and systemic bias of the anthropocentric tradition, it does not, in the final analysis, do much better. Although animal rights philosophy ostensibly affirms the conviction that, as Singer (1976) once described it, "all animals are equal," it turns out that "some animals," as George Orwell (1993, 88) put it, "are more equal than others." For leading animal rights theorists like Tom Regan, for example, only some animals—mainly cute and fuzzy-looking mammals of one year or more—qualify as morally significant. Other kinds of entities (e.g., reptiles, shellfish, insects, microbes) are practically insignificant, not worth serious moral consideration, and not even considered "animals" according to Regan's particular characterization of the term. This determination is not only prejudicial but morally suspect insofar as what Regan advocates—namely a critique of "Descartes's Denial," which effectively excludes animals from ethics—appears to be contradicted and undermined by what he does—marginalizing the majority of animal organisms from moral consideration.

In addition to instituting these other forms of segregation, animal rights philosophy is also unable or unwilling to consider the machine question. Although the animal and machine share a common form of alterity insofar as they are, beginning with Descartes's figure of the *animal-machine*, otherwise than human, only one of the pair has been granted access to consideration. The other of the animal other remains excluded and on the

periphery. Animal rights philosophy, therefore, only goes halfway in challenging the Cartesian legacy. And to add what is arguably insult to injury,
when Regan and other animal rights advocates make decisions about what
kind of animals to include and which organisms to exclude, they typically
do so by leveraging the very Cartesian strategy they contest, describing
these excluded others as mere mechanisms. The machine, therefore, continues to be the principal mechanism of exclusion, and this remains
unchallenged from the time of Descartes's *bête-machine* to Regan's *The Case
for Animal Rights*.

This problem, of course, has not gone unnoticed, and it is taken up and
addressed by developments in environmental ethics. As Sagoff (1984),
Taylor (1986), and Calarco (2008) have pointed out, the inherent difficulties of animal rights philosophy are clearly evident in the way that these
efforts exclude, and cannot help but exclude, all kinds of other subjects
from moral consideration. Toward this end, environmental ethics has
sought to provide for a more inclusive form of "universal consideration,"
where nearly anything and everything is a center of legitimate moral
concern. Although mainstream environmental ethics has tended to resist
extending this gesture of inclusivity in the direction of technological artifacts (see Taylor 1986), Luciano Floridi's proposal for information ethics
(IE) provides an elaboration that does appear to be able to achieve this
kind of universality. IE promises to bring to fulfillment the innovation that
began with efforts to address the animal question. Whereas animal rights
philosophy shifted the focus from a human-centered ethics to an animo-
centered system and environmental ethics took this one step further by
formulating a bio- or even ecocentric approach, IE completes the progres-
sion by advancing an ontocentric ethics that excludes nothing and can
accommodate anything and everything that has existed, is existing, or is
able to exist.

This all-encompassing totalizing effort is simultaneously IE's greatest
achievement and a significant problem. It is an achievement insofar as it
carries through to completion the patient-oriented approach that begins
to gain momentum with animal rights philosophy. IE promises, as Floridi
(1999) describes it, to articulate an "ontocentric, patient-oriented, ecological macroethics" that includes everything, does not make other problematic exclusions, and is sufficiently universal, complete, and consistent. It
is a problem insofar as this approach continues to deploy and support a

strategy that is itself part and parcel of a totalizing, imperialist program. The problem, then, is not which centrism one develops and patronizes or which form of centrism is more or less inclusive of others; the problem is with the centrist approach itself. All such efforts, as Lucas Introna (2009, 405) points out, "be it egocentric, anthropocentric, biocentric (Goodpaster 1978, Singer 1975) or even ecocentric (Leopold 1966, Naess 1995)—will fail." For all its promise, then, IE continues to employ a strategy of totalizing comprehension, whereby whatever is other is reduced to some common denominator by progressively lowering the level of abstraction so that what had been different can come to be incorporated within the community of the same. And that, of course, is the problem. What is needed, therefore, is another approach, one that does not continue to pursue a project of totalizing and potentially violent assimilation, one that is no longer satisfied with being merely revolutionary in its innovations, and one that can respond to and take responsibility for what remains in excess of the entire conceptual field that has been delimited and defined by the figures of agent and patient. What is needed is some way of proceeding and thinking otherwise.

3   Thinking Otherwise

3.1   Introduction

Moral philosophy has typically, in one way or another, made exclusive decisions about who is and who is not a legitimate moral agent and/or patient. We have, in effect, sought to determine the line dividing who or what is considered a member of the community of moral subjects from who or what remains outside. And we have done so by, as Thomas Birch (1993, 315) explains, assuming "that we can and ought to find, formulate, establish, institute in our practices, a criterion for (a proof schema of) membership in the class of beings that are moral *consideranda*." It is, for example, no longer news that many if not most of the Western traditions of ethics have been and, in many cases, continue to be exclusively anthropocentric. At the center of mainstream Western ethical theories—irrespective of the different varieties and styles that have appeared under names like virtue ethics, consequentialism, deontologism, or care ethics—has been a common assumption and virtually unquestioned validation of the άνθρωπος (*anthropos*)—the άνθρωπος who bears a responsibility only to other beings who are like him- or herself, that is, those who are also and already members of the community of άνθρωποι. This operation, from one perspective, is entirely understandable and even justifiable insofar as ethical theory is not some transcendent Platonic form that falls fully realized from the heavens but is rather the product of a particular group of individuals, made at a specific time, and initiated in order to protect a particular set of interests. At the same time, however, these decisions have had devastating consequences for others. In other words, any and all attempts to define and determine the proper limit of *consideranda* inevitably proceed by excluding others from participation. "When it comes to moral considerability," Birch (1993, 315) explains, "there *are*, and *ought* to be, insiders and outsiders, citizens and non-citizens (for example, slaves, barbarians, and women), 'members of the club' of *consideranda* versus the rest."

Ethics, therefore, has been and remains an exclusive undertaking. This exclusivity is fundamental, structural, and systemic. It is not accidental, contingent, or prejudicial in the usual sense of those words. And it is for this reason that little or nothing actually changes as moral theory and practices have developed and matured over time. Even when membership in the club of *consideranda* has, slowly and not without considerable resistance and struggle, been extended to some of these previously excluded others, there have remained other, apparently more fundamental and necessary exclusions. Or to put it another way, every new seemingly progressive inclusion has been made at the expense of others, who are necessarily excluded in the process. Animal rights philosophy, for instance, not only challenged the anthropocentric tradition in ethics but redefined the club of *consideranda* by taking a distinctly animo-centric approach where the qualifying criteria for inclusion in the community of moral subjects was not determined by some list of indeterminate humanlike capabilities—consciousness, rationality, free will, and so on—but the capacity to suffer, or "the ability to not be able," as Derrida (2008, 28) characterizes it.

This effort, despite its important innovations, still excludes others, most notably those nonmammalian animals situated on the lower rungs of the evolutionary ladder; other living organisms like plants and microbes; nonliving natural objects including soils, rocks, and the natural environment taken as a whole; and all forms of nonnatural artifacts, technologies, and machines. And even when these excluded others—these other kinds of others—are finally admitted into the club by other "more inclusive" lists of qualifying criteria that have been proposed by other moral theories, like environmental ethics, machine ethics, or information ethics, exclusions remain. There is, it seems, always someone or something that is and must be *other*. The sequence appears to be infinite, or what Hegel (1987, 137) termed "a bad or negative infinity": "Something becomes an other; this other is itself something; therefore it likewise becomes an other, and so on *ad infinitum*" (ibid.). Ethics, therefore, appears to be unable to do without its others—not only the others who it eventually comes to recognize as Other but also those other others who remain excluded, exterior, and marginalized. In the final analysis, ethics has been and continues to operate on the basis of a *fraternal logic*—one that defines and defends its membership by always and necessarily excluding others from participation in its exclusive and gated community.

Exclusion is certainly a problem. But inclusion, as its mere flip side and dialectical other, appears to be no less problematic. Despite the recent political and intellectual cachet that has accrued to the word, "inclusion" is not without significant ethical complications and consequences. "The inclusion of the other," as Jürgen Habermas (1998) calls it, whether another human being, animals, the environment, machines, or something else entirely, always and inevitably runs up against the same methodological difficulties, namely the reduction of difference to the same. In order to extend the boundaries of moral agency and/or patiency to traditionally marginalized others, philosophers have argued for progressively more inclusive definitions of what qualifies someone or something for ethical consideration. "The question of considerability has been cast," as Birch (1993, 314) explains by way of Kenneth Goodpaster, "and is still widely understood, in terms of a need for necessary and sufficient conditions which mandate practical respect for whomever or what ever fulfills them." The anthropocentric theories, for example, situate the human at the center of ethics and admit into consideration anyone who is able to meet the basic criteria of what has been decided to constitute the human being— even if, it should be recalled, this criterion has itself been something that is arguably capricious and not entirely consistent. Animal rights philosophy focuses attention on the animal and extends consideration to any organism that meets its defining criterion of "can they suffer?" The biocentric efforts of some forms of environmental ethics go one step further in the process, defining life as the common denominator and admitting into consideration anything and everything that can be said to be alive. And ontocentrism completes the expansion of moral consideration by incorporating anything that actually exists, had existed, or potentially exists, and in this way, as Floridi (1999) claims, provides the most universal and totalizing form of an all-inclusive ethics.

All these innovations, despite their differences in focus and scope, employ a similar maneuver. That is, they redefine the center of moral consideration in order to describe progressively larger and more inclusive circles that are able to encompass a wider range of possible participants.

Although there are and will continue to be considerable disagreements about who or what should define the center and who or what is or is not included, this debate is not the problem. The problem rests in the strategy itself. In taking this particular approach, these different ethical theories endeavor to identify what is essentially the same in a phenomenal diversity of individuals. Consequently, they include others by effectively stripping away and reducing differences. This approach, although having the appearance of being increasingly more inclusive, "is rather clearly a function of imperial power mongering," as Birch (1993, 315) describes it. For it immediately effaces the unique alterity of others and turns them into more of the same, instituting what Slavoj Žižek (1997, 161) calls the structure of the Möbius band: "At the very heart of Otherness, we encounter the other side of the Same." In making this argument, however, it should be noted that the criticism has itself employed what it criticizes. (Or to put it another way, the articulation of what is the matter is itself already and unavoidably involved with the material of its articulation.) In focusing attention on what is essentially the same in these various forms of moral centrism, the analysis does exactly what it charges—it identifies a common feature that underlies apparent diversity and effectively reduces a multiplicity of differences to what is the same. Pointing this out, however, does not invalidate the conclusion but demonstrates, not only in what is said but also in what is done, the questionable operations that are already involved in any attempt at articulating inclusion.

Exclusion is a problem because it calls attention to and fixates on what is different despite what might be similar. Inclusion is a problem, because it emphasizes similarities at the expense of differences. Consequently, the one is the inverse of the other. They are, as Michael Heim (1998, 42) calls them, "binary brothers," or, to put it in colloquial terms, two sides of one coin. As long as moral debate and innovation remain involved with and structured by these two possibilities, little or nothing will change. Exclusion will continue to be identified and challenged, as it has been in the discourses of moral personhood, animal rights, bioethics, and information ethics, by calls for greater inclusiveness and ethical theories that are able to accommodate these previously excluded others. At the same time, efforts to articulate inclusion will be challenged, as they have been in critical responses to these projects, as "imperial power mongering" (Birch 1993, 315) and for the reduction of difference that they had sought to respect and accommodate in the first place.

What is needed, therefore, is a third alternative that does not simply oppose exclusion by inclusion or vice versa. What is needed is an approach that is situated and oriented otherwise. In thinking otherwise, we will not be interested in taking sides or playing by the existing rules of the game. Instead we will be concerned with challenging, criticizing, and even changing the terms and conditions by which this debate has been organized, articulated, and configured. Precedent for this kind of alternative transaction can already be found in both the continental and analytic traditions. It is, for example, what poststructuralists like Jacques Derrida propose with the term "deconstruction," and what Thomas Kuhn endeavors to articulate in his paradigm changing work *The Structure of Scientific Revolutions*. But the practice is much older. It is, for instance, evident in Immanuel Kant's "Copernican revolution," which sought to resolve fundamental questions in modern philosophy not by lending support to or endeavoring to prove one or the other side in the rationalist versus empiricist debate but by rewriting rules of the game (Kant 1965, Bxvi).

But thinking otherwise is even older than this innovation in modern European thought, having its proper origin in the inaugural gesture attributed to the first philosopher, Socrates. It is, as John Sallis points out, in Plato's *Phaedo*, a dialogue that narrates among other things the final hours of Socrates's life, that the aged philosopher remembers where it all began. "In the face of death Socrates recalls how he became what he is: how he began by following the ways of his predecessors, the ways handed down through the operation of a certain tradition, the way of that kind of wisdom called περί φύσεως ίστορία; how this alleged wisdom repeatedly left him adrift; and how, finally taking to the oars, he set out on a second voyage by having recourse to λόγοι" (Sallis 1987, 1). As long as Socrates followed the tradition he had inherited from his predecessors—asking the questions they had already determined to be important, following the methods they had prescribed as being the most effective, and evaluating the kind of evidence they would recognize as appropriate—he failed. Rather than continue on this arguably fruitless path, he decides to change course by proceeding and thinking otherwise.

3.2 Decentering the Subject

Although not a Platonist by any means or even a philosopher, the roboticist Rodney Brooks supplies one effective method for pursuing the alternative of thinking otherwise. In *Flesh and Machines: How Robots Will Change Us*, Brooks describes, by way of an autobiographical gesture that is not unlike the one Plato has Socrates deploy, the "research heuristic" that was largely responsible for much of his success: "During my earlier years as a postdoc at MIT, and as a junior faculty member at Stanford, I had developed a heuristic in carrying out research. I would look at how everyone else was tackling a certain problem and find the core central thing that they had all agreed on so much that they never even talked about it. Then I would negate the central implicit belief and see where it led. This often turned out to be quite useful" (Brooks 2002, 37).

Following this procedure, we can say that one of the common and often unacknowledged features of the different formulations of both moral agency and patiency is the assumption that moral considerability is something that can and should be decided on the basis of individual qualities. This core assumption is clearly operational in, for example, the question of moral personhood, where the principal objective is to identify or articulate "the person-making qualities" in a way that is nonarbitrary and nonprejudicial; demonstrate how something, say, an animal or a machine, does in fact provide evidence of possessing that particular set of qualities; and establish guidelines that specify how such persons should be treated by others in the group. Even though there remain considerable disagreements about the exact qualities or criteria that should apply, what is not debated is the fact that an individual, in order to be considered a legitimate moral person, would need to achieve and demonstrate possession of the necessary and sufficient conditions for inclusion in the club. Instead of continuing in this fashion, arguing that some other individuals also clear the bar or making a case to revise the criteria of inclusion, we can proceed otherwise. Specifically, we can challenge or "negate," which is Brook's term, the basic assumption concerning the privileged place of the individual moral subject, arguably a product of Cartesian philosophy and the enlightenment's obsession with the self, with a decentered and distributed understanding of moral subjectivity. We can, in effect, agree that the center always and already cannot hold and that "things fall apart" (Yeats 1922, 289; Achebe 1994).

One such alternative can be found in what F. Allan Hanson (2009, 91) calls "extended agency theory," which is itself a kind of extension of actor-network approaches. According to Hanson, who takes what appears to be a practical and entirely pragmatic view of things, machine responsibility is still undecided and, for that reason, one should be careful not to go too far in speculating about the issue: "Possible future development of automated systems and new ways of thinking about responsibility will spawn plausible arguments for the moral responsibility of non-human agents. For the present, however, questions about the mental qualities of robots and computers make it unwise to go this far" (ibid., 94). Instead, Hanson, following the work of Peter-Paul Verbeek (2009), suggests that this problem may be resolved by considering various theories of "joint responsibility," where "moral agency is distributed over both human and technological artifacts" (Hanson 2009, 94). This is an elaboration of the "many hands" concept that had been proposed by Helen Nissenbaum (1996) to describe the distributed nature of accountability in computerized society.

In this way, Hanson's "extended agency" introduces a kind of "cyborg moral subject" where responsibility resides not in a predefined ethical individual but in a network of relations situated between human individuals and others, including machines. For Hanson, this distributed moral responsibility moves away from the anthropocentric individualism of enlightenment thought, which divides the world into self and other, and introduces an ethics that is more in line with recent innovations in ecological thinking:

When the subject is perceived more as a verb than a noun—a way of combining different entities in different ways to engage in various activities—the distinction between Self and Other loses both clarity and significance. When human individuals realize that they do not act alone but together with other people and things in extended agencies, they are more likely to appreciate the mutual dependency of all the participants for their common well-being. The notion of joint responsibility associated with this frame of mind is more conducive than moral individualism to constructive engagement with other people, with technology, and with the environment in general. (Hanson 2009, 98)

A similar proposal is provided in David F. Channell's *The Vital Machine*. After a rather involved investigation of the collapse of the conceptual differences customarily situated between technology and organisms, Channell ends his analysis with a brief consideration of "ethics in the age of the vital machine." "No longer," Channell (1991, 146) argues, beginning with a characterization that deliberately negates the standard approach, "can the focus of the theory of ethics be the autonomous individual. No longer can ethical judgments be based on a simple distinction between the intrinsic value of human beings and the instrumental value of technological creations. The focus of the ethics of the vital machine must be decentered." This decentering, however, does not go for all machines in all circumstances. Instead, it is contextualized in such a way as to be responsive to and responsible for differences in particular situations: "In some cases, with the use of traditional tools, the interactions may be very simple and the 'center' of ethics will be more on the side of the human, but in other cases, with the use of intelligent computers, the interactions may be quite complex and the 'center' of ethics will be more or less equally divided between the human and the machine" (ibid.). To respond to these apparent shifts in the 'center' of moral consideration, Channell proposes "a decentered ethical framework that reflects a bionic world view" (ibid., 152), what he calls "a bionic ethic" (ibid., 151).

This idea is derived from a reworking of Aldo Leopold's (1966) "land ethic." "While the land ethic of Leopold focuses on the organic, and in fact is usually interpreted as being in opposition to technology, it does provide a model for including both the organic and the mechanical into the expanding boundaries of a new ethic. In point of fact, Leopold often explained the interdependence of the biotic elements of nature in terms of engine parts or wheels and cogs" (Channell 1991, 153). Although often distinguished from technological concerns, Channell finds Leopold's land ethic to provide articulation of a moral thinking that can respect and take responsibility for nonliving objects, not only soils, waters, and rocks but also computers and other technological artifacts. For Channell, connecting the dots between these different concerns is not only a matter of metaphorical comparison—that is, the fact that nature has often been described and characterized in explicit mechanical terms—but grounded in established moral and legal precedent, that is, in the fact that "inanimate objects such as trusts, corporations, banks, and ships have long been seen by the courts as possessing rights"; the fact that some "writers have suggested that landmark buildings should be treated in a way similar to endangered species"; and the fact that "objects of artistic creation . . . have an intrinsic right to exist and be treated with respect" (ibid.).

In taking this approach, however, Channell is not arguing that inanimate objects and artifacts, like machines, should be considered the same as human beings, animals, or other living organisms. Instead, following

Leopold, he advocates a holistic ecological perspective, something that is called, borrowing a term from Richard Brautigan, "a cybernetic ecology." "The idea of a cybernetic ecology," Channell argues, "does not imply that machines should be given equal standing with humans or with animals, plants, rivers, or mountains. Even within nature, there is a hierarchy of living things, with some species dominant over others. A fundamental element of any ecological system is the 'food chain.' Throughout the environment the continued survival of one species is dependent on its being able to eat (or in more general terms transfer energy from) another part of the environment" (ibid., 154). The main issue, therefore, is figuring out where the various technological artifacts fit in the "food chain" of this "cybernetic ecology." Although this is, for now at least, still a largely undecided issue, what Channell proposes is a much more holistic approach and understanding of the moral landscape. For him, the issue is not simply who is and who is not part of an exclusive club, but rather how the different elements of the ecology fit together and support each other in a system that includes not just "deers and pines" but also "computers and electronics" (ibid.). "Within an ecological system, all elements have some intrinsic value but because of the interdependence within the system every element also has some instrumental value for the rest of the system. Each part of the ecology has a certain degree of autonomy, but in the context of the system, each part plays some role in the control of the entire ecology" (ibid.). What Channell advocates, therefore, is a shift in perspective from a myopic Cartesian subject to a holistic ecological orientation in which each element becomes what it is as a product of the position it occupies within the whole and is granted appropriate rights and responsibilities in accordance with the functioning and continued success of the entire system.

This decentered, systems approach to deciding questions of moral considerability sounds promising, but it has problems. First, this new cybernetic holism (and it should be recalled that cybernetics, from the very beginning, was to have been a totalizing science covering both animals and machines), which leverages the land ethic of Leopold, inherits many of the problems typically associated with environmental ethics. Although it challenges and deposes the anthropocentric privilege that human beings had traditionally granted themselves in moral philosophy, it still, as Birch (1993, 315) points out, locates a center of moral concern and organizes and regulates its system of ethics according to this new moral subject. Despite significantly challenging the anthropocentric perspective, this shift in focus is still and cannot help but be centrist. It simply redistributes what is considered to be the center of the moral universe. So for all its promise to decenter things, Channell's bionic ethic is just one more in a long line of competing and more inclusive forms of centrisms. Like Floridi's IE, it is clearly more universal and more inclusive, but it is, on this account, just more of the same.

Second, there is, in all of this, a problem with subjectivity. This comes out in the final paragraph of Channell's text, where he ends on an arguably optimistic if not utopian note: "In a cybernetic ecology both technology and organic life must be intelligently conserved. The entire system might be worse off without the peregrine falcon or the snail darter, but it also might be worse off without telecommunications and much of medical technology. On the other hand we might not want to conserve nuclear weapons or dioxin, but we might also be better off if the AIDS virus became extinct. In the end, we will build a new Jerusalem only if we can find a harmony between organic life and technology" (Channell 1991, 154). What remains unanswered in this optimistic assessment is: Who or what is the subject of this passage? Who or what is marked with the pronoun "we?" Who or what speaks in this conclusive statement? If the first-person plural refers to human beings, if it addresses itself to those individual humans who read the text and share a certain language, community, and tradition, then this statement, for all its promise, seems to sneak anthropocentrism in the back door. In this way, the cybernetic ecological perspective would become just another way of articulating, preserving, and protecting what are, in the final analysis, human interests and assets. And this conclusion seems to be supported by the examples Channell provides, insofar as the AIDS virus is something that adversely affects the immune system and integrity of the human species.

It is also possible, however, that this "we" refers not to human beings but to the vital machine and the entire cybernetic ecology. But then the issue must be who or what gives Channell, presumably a human being, the right to speak on behalf of this larger community. By what right does this individual, or anyone else, for that matter, write and speak on behalf of all the members of this community—human, animal, machine, or otherwise? Who or what grants this authority to speak, in this particular idiom, on behalf of this larger whole? This is of course the problem with any form of religious discourse, in which a particular human individual, like a prophet, or group of human beings, like a church, speaks on behalf of the divine, articulating what god wants, needs, or desires. Doing so is clearly expedient, but it is also a decisive imposition of power—what Birch (1993, 315) calls "imperial power mongering." Consequently, if Channell's "we" is human, it is not enough. If, however, his use of this term refers to the vital machine or the whole cybernetic ecology, it is perhaps too much. In pointing this out, my goal is not to fault Channell for getting it wrong but to point out how trying to get it right is already constrained and limited by the very system against which one struggles. Proposing an alternative, therefore, is neither simple, complete, nor beyond additional critical reflection.

An alternative thinking of decentered ethics is proposed by Joanna Zylinska (2009, 163), who advocates "a Deleuzian-influenced notion of 'distributive and composite agency.'" This form of "distributive agency" is proposed in direct response to and as an alternative for contemporary metaethical innovations that, although critical of the anthropocentric tradition, unfortunately do not go far enough. Zylinska, for instance, argues that the apparently groundbreaking work of animal rights philosophers, like Peter Singer, succeeds in "radicalizing humanist ethics by shifting the boundaries of who counts as a 'person'" while it "still preserves the structural principles of this ethics, with an individual person serving as the cornerstone" (ibid., 14). According to Zylinska, therefore, Singer merely remixes and modifies traditional anthropocentric ethics. He questions who or what gets to participate but ultimately preserves the basic structure and essential rules of the humanist game. In contrast, a concept of "distributive agency" recognizes and affirms the fact that an "individual person," however that comes to be defined, is always situated in and already operating through complex networks of interacting relations.

"Human agency," Zylinska argues with reference to the cyborg performance art of Stelarc, "does not disappear altogether from this zone of creative and contingent evolution, but it is distributed throughout a system of forces, institutions, bodies, and nodal points. This acknowledgement of agential distribution—a paradox that requires a temporarily rational and self-present self which is to undertake this realization—allows for an enactment of a more hospitable relationship to technology than the paranoid fear of the alien other" (ibid., 172). In this way, then, "distributive and composite agency" or what Zylinska also calls an "agency of assemblages" (ibid., 163) goes beyond the "many hands" thesis of Nissenbaun, Hanson's "extended agency theory," or Channell's "bionic ethic." Whereas these other theorists advocate the decentering of agency within a network of actors, Zylinska uses this distribution as a way to develop a distinctly Levinasian-inspired form of hospitality for others—one that can remain open to a completely different and alien other. In other words, where other forms of a decentered ethics inevitably focus attention on some other center—relying on the very structural gesture and approach that they had wanted to contest in the first place—Zylinska proposes a radical decentering in which nothing is a center of moral concern but everything can potentially be subject to ethics. What makes Zylinska's decentering work is its attention to the exorbitant other and other forms of otherness. This alternative way of looking at things ultimately concerns and will need to be referred to a reformulation of the question of moral patiency. In fact, this alternative approach seems to make patiency the privileged term rather than a derived aspect of some predefined notion of moral agency. Someone or something becomes a moral agent only after first being admitted into the fraternity of moral subjects—only after and on the condition that some other is recognized as Other.

3.3   The Ethics of Social Construction

A decentered approach recognizes the way moral responsibility is often not constituted by an individual subject but instead comes to be distributed across a network of interrelated and interacting participants. But this is not the only way to proceed. Other alternatives focus not on a decentering of the moral subject by tracing its distribution within the social fabric of a network but consider how the moral subject, whether conceptualized as an agent, patient, or both, has been socially constructed, regulated, and assigned. One such alternative is advanced by Bernd Carsten Stahl under the concept of "quasi-responsibility." In the article "Responsible Computers? A Case for Ascribing Quasi-Responsibility to Computers Independent of Personhood or Agency," Stahl effectively skirts the question of agency and personhood by reformulating the entire approach: "Instead of engaging in questions of agency or personhood and the analysis of when computers can become subjects of (moral) responsibility, [this] paper introduces a different type of responsibility. This quasi-responsibility encompasses only a limited sub-set of traditional responsibility but it is explicitly applicable to non-human subjects, including computers" (Stahl 2006, 212).

Instead of trying to answer or even give serious attention to the seemingly irresolvable question concerning machine moral agency and personhood, Stahl, following the precedent and strategy modeled by Alan Turing, changes the question by limiting inquiry to "quasi-responsibility." This term, which Stahl recognizes is not very elegant, "follows Ricoeur who suggested a 'quasi-agency' for historical collectives such as states or nations who can be described usefully as agents even though they are not traditional agents" (ibid., 210). Accordingly, "quasi-responsibility" is characterized as a socially constructed attribution of agency that takes place independent of any traditional consideration of agency or personhood:

The term "quasi-responsibility" indicates that the speaker intends to use the idea of a social construction for the purpose of ascribing a subject to an object with the aim of attributing sanctions (the heart of responsibility) without regard to the question of whether the subject fulfills the traditional conditions of responsibility. It shows that the focus of the ascription is on the social outcomes and consequences, not on considerations of agency or personhood. The concept was developed using computers as a main example but there is no fundamental reason why it could not be extended to other non-human entities, including animals. (Ibid.)

Stahl, therefore, advocates an alternative conceptualization of moral responsibility that is "a social construct of ascription" (ibid.), completely disengaged from and not concerned with the customary debates about and questions of moral agency and personhood.

Anne Foerst takes this innovation one step further, by contesting the location of agency and the way it is (or is not) assigned, and she does so by revisiting and significantly revising the concept of person. "'Person,'" Foerst explains in an interview, "is an assignment, given to each one of us by our parents and our closest community right after birth. It is given to us by God in the first place, and we are free to assign it to others. But we are also free to deny it to others" (Benford and Malartre 2007, 162–163). According to Foerst, the defining feature of personhood is not something discovered within the metaphysical fabric or psychological makeup of an individual being. It is not something that individuals possess as their personal property and then can exhibit to others in some kind of test or demonstration, which would presumably be grounds for either including them in or excluding them from the community of moral subjects. Instead, the concept "person" is a social construct and assignment that is conferred (or not) by others through explicit decision. In this way, then, the question of personhood, as Hilary Putnam (1964, 691) wrote with regard to the question of artificial life, "calls for a decision and not a discovery." And this is, Foerst argues, how things actually work: "Each of us only assigns personhood to a very few people. The ethical stance is always that we have to assign personhood to everyone, but in reality we don't. We don't care about a million people dying in China of an earthquake, ultimately, in an emotional way. We try to, but we can't really, because we don't share the same physical space. It might be much more important for us if our dog is sick" (Benford and Malartre 2007, 163).

When asked directly whether this subject position can be legitimately assigned to machines, Foerst answers in the affirmative: "I think a machine can definitely be a person. The more social our machines get, the more interactive they are, the more they learn out of interaction with us, the creators, and the more we interact with them. For me there is no question that at some point they will be persons like anyone else. They will not be humans because they are different material, but they will be part of our community, there is no question in my mind" (ibid.). According to Foerst's argument, one is not first defined as a moral person who then subsequently engages in actions with others. Instead one is assigned the position of moral person (whether that be an agent, patient, or both) as a product of social relations that precede and prescribe who or what one is. And a machine can come to occupy this particular subject position just as easily as a human being, an animal, an organization, and so on.

This substantially reconfigures the field of inquiry. "What we should be arguing about," R. G. A. Dolby writes in "The Possibility of Computers Becoming Persons," "is not the possibility of machine souls or machine minds, but whether robots could ever join human society. The requirement that must be met by a robot is that people are prepared to treat it as a person. If they are, they will also be prepared to attribute to it whatever inner qualities they believe a person must have" (Dolby 1989, 321). Consequently, personhood is not decided on the basis of the possession of some occult metaphysical properties that remain for all our efforts beyond epistemological grasp but is something that is socially constructed, negoti-ated, and conferred. It is not the case that someone or something first shows evidence of possessing the predefined qualities of moral subjectivity and then subsequently engages in intersubjective relationships that are determined to be more or less morally correct. The order of precedence is reversed. Someone or something that is already engaged in some particular interaction with others is responded to and accorded the status of moral subjectivity in and by the process of the interaction. The moral *subject* is literally thrown behind or under the interaction as the source and support of the interaction from which it is initially derived. In this way, the moral subject is something that is "(presup)posited" (Žižek 2008a, 209).

This means that the moral person is not some predefined, stable, and well-established ontological position but is, as Dennett (1998, 285) describes it, a "normative ideal." In other words, "the concept of a person is only a free-floating honorific that we are all happy to apply to ourselves, and to others as the spirit moves us, guided by our emotions, aesthetic sensibili-ties, considerations of policy, and the like" (ibid., 268). Understood in this fashion, moral personhood does not require a Turing test or similar dem-onstration. All that is necessary is evidence that someone or something has been (for whatever reason) considered a person by others within a particular community. This is precisely the argumentative approach employed by Leiber's fictional complainant: "The arguable question is whether they [the chimpanzee Washoe-Delta and the computer AL] are indeed persons and hence we should begin with that. We say that Washoe-Delta and AL interacted with, and were recognized by, the human crew as persons" (Leiber 1985, 5). To be a person, in other words, it is enough that one be addressed and treated by others as a person. Or, to borrow a statement from Donna Haraway (2008, 17), "the partners do not precede their relating" but become who and what they are in and through the relationship.

According to the alternative approaches of Foerst, Dolby, and Stahl, someone or something becomes a moral subject with legitimate ethical standing not on the prior determination and demonstration of his/her/its agency or the possession of some psychological properties that are consid-ered to be "person-making" but by being situated, treated, and responded to as another person by a particular community in concrete situations and encounters. This means that "person" is not, as Foerst concludes, some

"empirical fact"; instead it is a dynamic and socially constructed hono-
rarium or "gift" (Benford and Malartre 2007, 165) that can be bestowed
(or not) on others by a particular community in a particular place for a
particular time. For this reason, "person" (assuming it is decided to retain
this word) is never something that is absolute and certain but is always
and already a relative term, the assignment of which has its own moral
implications and consequences.

This other approach, which situates the ethical relationship with others
prior to and not dependent on or derived from a predefined ontological
decision is promising. Despite its promise, however, these alternatives
are not without their critics. "There are so many oddities," Christopher
Cherry (1991, 22) writes in direct response to Dolby's article, "on so many
levels":

What can it mean to talk of machines "join[ing] human society," being "treated,"
"accepted," as persons, and so forth? Dolby thinks we should start (not end) by
awarding colours. We can then, but only if we feel the need, hang on them sentient
"qualities"—in my words, see sentient states in them. Here there are vulgar parallels
with the theologian's injunction: act in order to believe, and believe in order to
understand. At the best of times this is puzzling advice: on this occasion it is unap-
pealing as well, intimating that our very humanity is a metaphysical fancy to be
indulged. (Ibid.)

For Cherry, the alternative way of explaining personhood is not only
threatening to human dignity but "deeply incoherent" and philosophi-
cally deficient. Or as Jay Friedenberg (2008, 2) describes it, admittedly using
something of a caricature of the position, "the idea of a person being a
person because someone else thinks they are is unsatisfying from a scientific perspective. Science is an objective endeavor and would like to be able
to find some crucial human physical property that can be measured and
detected."

The real problem in all this, however, is that it just reverses things,
making moral personhood a product of the relationship and not the other
way around. Although a promising intervention, inversion is not enough;
it never is. In turning things around, it still operates according to a logic
that is oriented and controlled by the very system that is inverted. Instead
of being the origin of the relationship, the individual moral subject is now
its product. This approach, therefore, is still organized and regulated by a
centrist logic. Consequently, this innovation, although promising, is not

sufficient to break with the individualistic tradition; it is only its negative and inverted form. Although recognizing that the club of moral *consideranda* is socially constructed and that we are responsible for deciding on its membership, this method still operates on the grounds of a questionable fraternal logic. We are still organizing, to deploy a phrase popularized by Todd Solondz's *Welcome to the Dollhouse* (1995), "special people clubs" and deciding who is and who is not special. This may provide for a more honest understanding of the way these decisions are made, but what it does not account for is the fact that this decision is itself an ethical matter with its own assumptions and consequences. There is, then, a kind of morality before morality. Before something is decided to be either a moral agent or moral patient, we make a decision whether to make this decision or not. Or as Søren Kierkegaard (1987, 169) described this redoubled moral decision, "rather than designating the choice between good and evil, my Either/Or designates the choice by which one chooses good and evil or rules them out."

3.4 Another Alternative

When it comes to thinking otherwise about others, especially as it relates to the question concerning ethics, there is perhaps no scholar better suited to the task than Emmanuel Levinas. Unlike a lot of what goes by the name of "moral philosophy," Levinasian thought does not rely on metaphysical generalizations, abstract formulas, or simple pieties. Levinasian philosophy not only is critical of the traditional tropes and traps of Western ontology but proposes an ethics of radical otherness that deliberately resists and interrupts the metaphysical gesture par excellence, that is, the reduction of difference to the same. This radically different approach to thinking difference is not just a useful and expedient strategy. It is not, in other words, a mere gimmick. It constitutes a fundamental reorientation that effectively changes the rules of the game and the standard operating presumptions. In this way, "morality is," as Levinas (1969, 304) concludes, "not a branch of philosophy, but first philosophy." This fundamental reconfiguration, which puts ethics *first* in both sequence and status, permits Levinas to circumvent and deflect a lot of the difficulties that have traditionally tripped up moral thinking in general and efforts to address the machine question in particular.

First, for Levinas, the problems of other minds[1]—the seemingly irresolvable fact that one cannot know with any certitude whether the other who confronts me either has a conscious mind or not—is not some fundamental epistemological limitation that must be addressed and resolved prior to moral decision making but constitutes the very condition of the ethical relationship as an irreducible exposure to an Other who always and already exceeds the boundaries of the self's totalizing comprehension. Consequently Levinasian philosophy, instead of being derailed by the standard epistemological problem of other minds, immediately affirms and acknowledges it as the basic condition of possibility for ethics. Or as Richard Cohen succinctly describes it in what could be a marketing slogan for Levinasian thought, "not 'other minds,' mind you, but the 'face' of the other, and the faces of all others" (Cohen 2001, 336). In this way, then, Levinas provides for a seemingly more attentive and empirically grounded approach to the problem of other minds insofar as he explicitly acknowledges and endeavors to respond to and take responsibility for the original and irreducible difference of others instead of getting involved with and playing all kinds of speculative (and unfortunately wrongheaded) head games. "The ethical relationship," Levinas (1987, 56) writes, "is not grafted on to an antecedent relationship of cognition; it is a foundation and not a superstructure. . . . It is then more *cognitive* than cognition itself, and all objectivity must participate in it."

Second, and following from this, Levinas's concern with/for the Other will constitute neither an agent- nor patient-oriented ethics, but addresses itself to what is anterior to and remains in excess of this seemingly fundamental logical structure. Although Levinas's attention to and concern for others looks, from one perspective at least, to be a kind of "patient-oriented" ethics that puts the interests and rights of the Other before oneself, it is not and cannot be satisfied with simply endorsing one side of or conforming to the agent–patient dialectic. Unlike Floridi's IE, which advocates a patient-oriented ethic in opposition to the customary agent-oriented approaches that have maintained a controlling interest in the field, Levinas goes one step further, releasing what could be called a *deconstruction*[2] of the very conceptual order of agent and patient. This alternative, as Levinas (1987, 117) explains, is located "on the hither side of the act-passivity alternative" and, for that reason, significantly reconfigures the standard terms and conditions. "For the condition for," Levinas explains, "or the unconditionality of, the self does not begin in the auto-affection of a sovereign ego that would be, after the event, 'compassionate' for another. Quite the contrary: the uniqueness of the responsible ego is possible only in being obsessed by another, in the trauma suffered prior to any auto-identification, in an unrepresentable *before*" (ibid., 123). The self or the ego, as Levinas describes it, does not constitute some preexisting self-assured condition that is situated before and is the cause of the subsequent relationship with an other. It does not (yet) take the form of an active agent who is able to decide to extend him- or herself to others in a deliberate act of compassion. Rather it becomes what it is as a by-product of an uncontrolled and incomprehensible exposure to the face of the Other that takes place prior to any formulation of the self in terms of agency.

Likewise, the Other is not comprehended as a patient who is constituted as the recipient of the agent's actions and whose interests and rights would need to be identified, taken into account, and duly respected. Instead, the absolute and irreducible exposure to the Other is something that is anterior and exterior to these distinctions, not only remaining beyond the range of their conceptual grasp and regulation but also making possible and ordering the antagonistic structure that subsequently comes to characterize the difference that distinguishes the self from its others and the agent from the patient in the first place. In other words, for Levinas at least, prior determinations of agency and patiency do not first establish the terms and conditions of any and all possible encounters that the self might have with others and with other forms of otherness. It is the other way around. The Other first confronts, calls upon, and interrupts self-involvement and in the process determines the terms and conditions by which the standard roles of moral agent and moral patient come to be articulated and assigned. Consequently, Levinas's philosophy is not what is typically understood as an ethics, a metaethics, a normative ethics, or even an applied ethics. It is what John Llewelyn (1995, 4) has called a "proto-ethics" or what others have termed an "ethics of ethics." "It is true," Derrida explains, "that Ethics in Levinas's sense is an Ethics without law and without concept, which maintains its non-violent purity only before being determined as concepts and laws. This is not an objection: let us not forget that Levinas does not seek to propose laws or moral rules, does not seek to determine *a* morality, but rather the essence of the ethical relation in general. But as this determination does not offer itself as a *theory* of Ethics, in question, then, is an Ethics of Ethics" (Derrida 1978, 111).

Third, and because of this, Levinasian thought does not, like many of the other attempts to open up moral thinking to other forms of otherness, get hung up on debating and deciding the issue of moral personhood. The Other is not yet, strictly speaking, another person. In other words, I respond to and have a responsibility for the Other not because he/she/it is always already another person like I assume myself to be--a kind of alter ego with a similar set of properties, rights, and responsibilities--but because the Other is always and already otherwise. "Levinas," as Simmone Plourde (2001, 141) explains, "pushed the concept of personhood to its most intimate depths by substituting it with the notion of the face." But the face of the Other is not necessarily that of another "person" as that term has been deployed and developed in the history of moral philosophy.

Personhood is typically decided prior to the ethical relationship based on the articulation of an exclusive criteria and a decision as to who is and who does not possess the appropriate person-making qualities. In Kant's view, for instance, "the other person who is the object of my moral action is constituted *after* I have constituted myself as a moral subject a priori. This other person, then, is essentially an analogue of my own fully conscious moral personhood" (Novak 1998, 166). Conceptualized in this way, I have an ethical obligation to anyone who or anything that clears the bar for inclusion in the club of persons to which I already assume that I belong. All others who fail to measure up fall outside the scope of such obligations and may be dispensed with as I please and without further consideration. This is the case, whether personhood comes to be defined with a specific list of qualifying criteria (Cherry 1991; DeGrazia 2006; Dennett 1998; Scott 1990; Smith 2010) or delimited by deciding on an appropriate level of abstraction (Floridi and Sanders 2004). In whatever manner it comes to be determined, membership in the community of moral persons is something that is decided a priori, and the moral obligation is predicated upon and subsequent to this decision.

Levinas deliberately turns things around by reconfiguring the assumed order of precedence. For him, the ethical relationship, the moral obligation that the I has to the Other, precedes and determines who or what comes, after the fact, to be considered a moral subject or "person." Ethics, therefore, is not predicated on an a priori ontological determination of personhood. Instead, personhood, assuming that this word is to be retained, is something that is first determined on the basis of and as a product of the ethical relationship. "Modern antihumanism," Levinas (1987, 127–128) writes, "which denies the primacy that the human person, free for itself, would have for the signification of being, is true over and beyond the reasons it gives itself. It clears the place for subjectivity positing itself in abnegation, in sacrifice, in a substitution which precedes the will. Its inspired intuition is to have abandoned the idea of person, goal and origin of itself, in which the ego is still a thing because it is still a being." According to Levinas, therefore, the Other always and already obligates me in advance of the customary decisions and debates concerning personhood and who or what is and is not a moral subject. "If ethics arises," Calarco (2008, 71) writes, "from an encounter with an Other who is fundamentally irreducible to and unanticipated by my egoistic and cognitive machinations," then identifying the "'who' of the Other" is something that cannot be decided once and for all or with any certitude. This apparent inability or indecision, however, is not necessarily a problem. In fact, it is a considerable advantage insofar as it opens ethics not only to the Other but to other forms of otherness. "If this is indeed the case," Calarco concludes, "that is, if it is the case that we do not know where the face begins and ends, where moral considerability begins and ends, then we are obligated to proceed from the possibility that anything might take on a face. And we are further obligated to hold this possibility permanently open" (ibid.).

Levinasian philosophy, therefore, does not make prior commitments or decisions about who or what will be considered a legitimate moral subject. For Levinas, it seems, anything that faces the I and calls its *ipseity* into question would be Other and would constitute the site of ethics. Despite the promise this innovation has for arranging a moral philosophy that is radically situated otherwise, Levinas's work is not able to escape from the anthropocentric privilege. Whatever the import of his unique contribution, Other in Levinas is still and unapologetically human. Although he is not the first to identify it, Jeffrey Nealon provides what is perhaps one of the most succinct descriptions of this problem in *Alterity Politics*: "In thematizing response solely in terms of the human face and voice, it would seem that Levinas leaves untouched the oldest and perhaps most sinister unexamined privilege of the same: *anthropos* [ἄνθρωπος] and only *anthropos*, has *logos* [λόγος]; and as such, *anthropos* responds not to the barbarous or the inanimate, but only to those who qualify for the privilege of 'humanity,' only those deemed to possess a face, only to those recognized to be living in the *logos*" (Nealon 1998, 71). This residue of the human and of a certain brand of humanism is something that had also structured Derrida's critical response to Levinas's work in the introduction to the 1997 presentation at Cerisy-la-Salle and is the subject of Richard Cohen's introduction to the English translation of Levinas's 1972 *Humanism of the Other* (Levinas 2003).

For Derrida, the humanist pretensions of Levinasian philosophy constitute cause for considerable concern: "In looking at the gaze of the other, Levinas says, one must forget the color of his eyes, in other words see the gaze, the face that gazes before seeing the visible eyes of the other. But when he reminds us that the 'best way of meeting the Other is not even to notice the color of his eyes,' he is speaking of man, of one's fellow as man, kindred, brother; he thinks of the other man and this, for us, will later be revealed as a matter for serious concern" (Derrida 2008, 12). And what truly "concerns" Derrida is not just the way this anthropocentrism limits Levinas's philosophical innovations but the way it already makes exclusive decisions about the (im)possibility of an animal other. "Emmanuel Levinas did not," Derrida points out, "make the animal anything like a focus of interrogation within his work. This silence seems to me here, at least from the point of view that counts for us, more significant than all the differences that might separate Levinas from Descartes and from Kant on the question of the subject, of ethics, and of the person" (ibid., 105–106).

Additionally, on those one or two rare occasions when Levinas does address himself directly to animals—when he is, it seems, not silent about the animal—it is in order to silence or dismiss them from any further consideration. There is, for instance, the well-known situation with Bobby, a dog that Levinas and his fellow prisoners of war encountered during their incarceration by the Germans (Levinas 1990, 152–153). Although Levinas directly, and with what one might call considerable compassion, addresses himself to the relationship the prisoners had with this particular animal, whom Levinas nominates as "the last Kantian in Nazi Germany" (ibid., 153), he only includes the dog in order to marginalize him. Bobby has, Levinas points out, "neither ethics nor *logos*" (ibid., 152). And in making this distinction, Levinas's consideration of Bobby does not subvert or question but exhibits affiliation with the Cartesian tradition that Derrida charges him with: "Bobby is," as David Clark (2004, 66) points out, "thus closer to a cyborg than to a sentient creature; his is not unlike an empty machine of the sort Descartes hallucinated when he looked at animals."

Whereas Derrida maintains a critical stance toward Levinas's silence on the animal question and its rather conventional underlying humanism, Richard Cohen endeavors to give it a positive spin: "The three chapters of *Humanism of the Other* each defend humanism—the world view founded on the belief in the irreducible dignity of humans, a belief in the efficacy and worth of human freedom and hence also of human responsibility" (Cohen 2003, ix). For Cohen, however, this humanism is not the common run-of-the-mill anthropocentrism; it consists in a radical thinking of the human person as the unique site of ethics:

> From beginning to end, Levinas's thought is a humanism of the other. The distinctive moment of Levinas's philosophy transcends its articulation but is nevertheless not difficult to discern: the superlative moral priority of the other person. It proposes a conception of the "humanity of the human," the "subjectivity of the subject," according to which being "for-the-other" takes precedence over, is better than being for-itself. Ethics conceived as a metaphysical anthropology is therefore nothing less than "first philosophy." (Ibid., xxvi)

What is important to note in these two different responses to the anthropocentrism that is apparent in Levinas's work is the fact that both sides recognize and affirm a fundamental and irreducible form of humanism that is always and already at work within Levinas's ethics of otherness. For Levinas, as for many of those thinkers who came before and follow in the wake of his influence, the other is always and already operationalized as another human subject. For Derrida this is a matter for serious concern insofar as it threatens to undermine Levinas's entire philosophical enterprise. For Cohen, however, it is an indication of Levinas's unique dedication to and focus on the humanism of the Other. Either way, if, as Levinas argues, ethics precedes ontology, then in Levinas's own work, anthropology and a certain brand of humanism appear to precede and underwrite ethics.

3.4.1 The Animal Other

If Levinasian philosophy is to provide a way of thinking otherwise that is able to respond to and to take responsibility for other forms of otherness, or to consider and respond to, as John Sallis (2010, 88) describes it, "the question of another alterity," we will need to use and interpret Levinas's own philosophical innovations in excess of and in opposition to him. We will need, as Derrida (1978, 260) once wrote of Georges Bataille's exceedingly careful engagement with the thought of Hegel, to follow Levinas to the end, "to the point of agreeing with him against himself" and of wresting his discoveries from the limited interpretations that he provided. Such efforts at "radicalizing Levinas," as Peter Atterton and Matthew Calarco (2010) refer to it, will take up and pursue Levinas's moral innovations in excess of the rather restricted formulations that he and his advocates and critics have typically provided. Calarco, in particular, takes issue with the way Levinas or at least certain readings of his work have simply and without any significant critical hesitation excluded the animal from ethics. "The two dominant theses," Calarco (2008, 55) writes, "in Levinas's writings concerning animals are: no nonhuman animal is capable of genuine ethical response to the Other; nonhuman animals are not the kind of beings that elicit an ethical response in human beings—which is to say, the Other is always and only the *human* Other." As Calarco correctly points out, the customary reading of Levinas, one that is shared by many of his critics and advocates alike, Derrida included, is that he denies both moral agency and patiency to the animal. In other words, the animal is not able to respond to others, nor does it constitute an Other who calls for an ethical response.

For Calarco, however, this is not and should not be the final word on the matter: "Although Levinas himself is for the most part unabashedly and dogmatically anthropocentric, the underlying logic of his thought permits no such anthropocentrism. When read rigorously, the logic of Levinas's account of ethics does not allow for either of these two claims. In fact, as I shall argue, Levinas's ethical philosophy is, or at least should be, committed to a notion of *universal ethical consideration*, that is, an agnostic form of ethical consideration that has no a priori constraints or boundaries" (ibid.). In proposing this alternative approach, Calarco interprets Levinas against himself, arguing that the logic of Levinas's account of ethics is in fact richer and more radical than the limited interpretation that the philosopher had initially provided for it. Calarco, therefore, not only exposes and contests Levinas's anthropocentrism, which effectively sidelines from ethical consideration any and all nonhuman things, but seeks to locate in his writings the possibility for articulating something like Birch's (1993) "universal consideration." In fact, Calarco not only employs Birch's terminology but utilizes his essay as a way of radicalizing the Levinasian perspective: "Rather than trying to determine the definitive criterion or criteria of moral considerability, we might, following Birch and the reading of Levinas that I have been pursuing, begin from a notion of 'universal consideration' that takes seriously our fallibility in determining where the face begins and ends. Universal consideration would entail being ethically attentive and open to the possibility that anything might take on face" (Calarco 2008, 73).

This radical possibility obviously opens the door to what some might consider an absurd conclusion. "At this point," Calarco admits, "most reasonable readers will likely see the argument I have been making as having absurd consequences. While it might not be unreasonable to consider the possibility that 'higher' animals who are 'like' us, animals who have sophisticated cognitive and emotive functions, could have a moral claim on us, are we also to believe that 'lower' animals, insects, dirt, hair, fingernails, ecosystems and so on could have a claim on us?" (ibid., 71). In responding to this charge, Calarco deploys that distinctly Žižekian (2000, 2) strategy of "*fully endorsing what one is accused of.*" "I would suggest," Calarco (2008, 72) argues, "affirming and embracing what the critic sees as an absurdity. All attempts to shift or enlarge the scope of moral consideration are initially met with the same reactionary rejoinder of absurdity from those who uphold common sense. But any thought worthy of the name, especially any thought of ethics, takes its point of departure in setting up a critical relation to common sense and the established doxa and, as such, demands that we ponder absurd, unheard-of thoughts."

A similar decision is made by John Llewelyn, who recognizes that we always risk sliding down a slippery slope into "nonsense" when we attempt to take others and other forms of otherness into account:

We wanted to open the door of ethical considerability to animals, trees, and rocks. This led us to propose a distinction that promised to allow us to open the door in this way without thereby opening the door to the notion that we have ethical responsibilities to, for example, numbers. Is this notion a nonsense? Maybe. But nonsense itself is historically and geographically relative. What fails to make sense at one time and place makes sense at another. This is why ethics is educative. And this is why for the existentist concept of the ethical this chapter has been projecting to be as extensive and democratic as justice demands we may be ethically obligated to talk nonsense. (Llewelyn 2010, 110–111)

Talking nonsense was, of course, the case with the animal question, which was initially advanced by Thomas Taylor (1966) as a kind of absurdity in order to ridicule what was assumed to be another absurdity—the extension of rights to women. It is again encountered in Christopher Stone's (1974) consideration of the seemingly absurd and "unthinkable" question "Should Trees Have Standing?" "Throughout legal history," Stone (1974, 6–7) writes, "each successive extension of rights to some new entity has been, theretofore, a bit unthinkable. . . . The fact is, that each time there is a movement to confer rights onto some new 'entity,' the proposal is bound to sound odd or frightening or laughable." And it is with suppressed laughter—the kind of embarrassed laughter that is always on the verge of bursting forth in the face of what appears to be nonsense—that the apparent absurdity of the machine question has been given consideration, as is evident from the editors' note appended to the beginning of Marvin Minsky's "Alienable Rights": "Recently we heard some rumblings in normally sober academic circles about robot rights. We managed to keep a straight face as we asked Marvin Minsky, MIT's grand old man of artificial intelligence, to address the heady question" (Minsky 2006, 137).

Calarco's reworking of Levinasian philosophy, therefore, produces a much more inclusive ethics that is able to take other forms of otherness into account. And it is, no doubt, a compelling proposal. What is interesting about his argument, however, is not the other forms of otherness that come to be included through his innovative reworking of Levinas, but what (unfortunately) gets left out in the process. According to the letter of Calarco's text, the following entities should be given moral consideration: "'lower' animals, insects, dirt, hair, fingernails, and ecosystems." What is obviously missing from this list is anything that is not "natural," that is, any form of artifact. Consequently, what gets left behind or left out by Calarco's "universal consideration"—a mode of ethical concern that does not shrink from potential absurdities and the unthinkable—are tools, technologies, and machines. It is possible that these excluded others might be covered by the phrase "and so on," which Calarco appends to the end of his litany, in much the same way that "an embarrassed 'etc.,'" as Judith Butler (1990, 143) calls it, is often added to a string of others in order to gesture in the direction of those other others who did not make the list.

But if the "and so on" indicates, as it typically does, something like "more along same lines as what has been named," then it seem that the machine would not be included. Although Calarco (2008, 72) is clearly prepared, in the name of the other and other kinds of otherness, "to ponder absurd, unheard-of thoughts," the machine remains excluded and in excess of this effort, comprising a kind of absurdity beyond absurdity, the unthinkable of the unthought, or the other of all who are considered Other. According to Calarco, then, the resistance that is offered to *ipseity* by all kinds of other nonhuman things does, in fact, and counter to Levinas's own interpretation of things, make an ethical impact. But this does not apply, it seems, to machines, which remain, for both Levinas and Calarco, otherwise than Ethics, or beyond Other.

3.4.2 Other Things

A similar proposal, although formulated in an entirely different fashion, is advanced in Silvia Benso's *The Face of Things* (2000). Whereas Calarco radicalizes Levinas by way of Birch's "universal consideration," Benso's efforts to articulate "a different side of ethics" take the form of a forced confrontation between Levinasian ethics and Heideggerian ontology:

> The confrontation between Heidegger and Levinas, in their encounter with things, is marked by a double truth. On the one hand, there are no things in Levinas, since things are for the same, or for the Other, but not for themselves. That is, in Levinas there is no alterity of things. Conversely, there are things in Heidegger. For him, things are the place where the gathering of the Fourfold—the mortals, the gods, the earth, the sky—comes to pass, in an intimacy that is not fusion but differing. Each thing remains other in hosting the Fourfold in its peculiar way: other than the Fourfold and other than any other thing; other than the mortals, who can dwell by things in their thinging only if they can take care of things as things, if they can let them be in their alterity. . . . Undoubtedly there is ethics in Levinas, even if his notion of ethics extends only to the other person (certainly the other man, hopefully also the other woman and child). Conversely there is no ethics in Heidegger, at least according to the most common reading. If the two thinkers are forced face to face in a confrontation that neither of them would advocate enthusiastically, the result is a chiasmatic structure, whose branches connect a double negation—nonethics and nonthings—and a double affirmation—ethics and things. (Benso 2000, 127)

What Benso describes, even if she does not use the term, is a *mashup*. "Mashup" refers to a digital media practice where two or more recordings, publications, or data sources are intermingled and mixed together in order to produce a third term that is arguably greater than the sum of its constituent parts. Although the practice has recently proliferated across all forms of digital media content, becoming what William Gibson (2005, 118) has called "the characteristic pivot at the turn of our two centuries," it was initially deployed and developed in the field of popular music. Perhaps the best-known audio mashup is DJ Danger Mouse's *Grey Album*, which comprises a clever and rather unexpected combination of vocals taken from Jay-Z's *Black Album* layered on top of music extracted from one of the undisputed classics of classic rock, the Beatles' *White Album*. Benso does something similar with two philosophers who are at least as different as Jay-Z is from the Beatles, staging a confrontation between Levinas and Heidegger that neither thinker would want but which, irrespective of that, produces an interesting hybrid of the two.

What can be heard, seen, or read in this unauthorized remix of Levinasian ethics and Heideggerian ontology can, as Benso predicts, result in either a thinking of ethics and things or, its negative image, a thinking of nonethics and nonthings. In the face of these two possibilities, Benso suggests that the latter would not only be too easy but would result in a rather predictable outcome, which would inevitably sound like everything else. She, therefore, endeavors to take up and pursue the path less traveled. "Since Socrates," Benso (2000, 127–128) explains in what might be considered a Nietzschean mood,

philosophy has walked the path of negation. If there is ethics, it is not of things; and if there are things, they are not ethical. The path of affirmation is a narrow strip, which has seldom been explored. It leads to an ethics of things, where ethics cannot be traditional ethics in any of its formulations (utilitarian, deontological, virtue-oriented), and things cannot be traditional things (objects as opposed to a subject). At the intersection between ethics and things, Levinas and Heidegger meet, as in "a contact made in the heart of chiasmus." The former offers the notion of a nontraditional ethics, the latter of nontraditional things.

Benso's philosophical remix sounds both inventive and promising. Like Calarco's "universal consideration," it endeavors to articulate a distinctly Levinasian ethics that is no longer exclusively anthropo- or even biocentric—one that can, in fact, accommodate itself to and is able to respond appropriately in *the face of things*. In addition to this, the mashup of Levinas's "nontraditional ethics" and Heidegger's "nontraditional things" does some important philosophical heavy lifting, facilitating a meeting between the two parties where the one critiques and *supplements* the other in a way that would not be simply "critical" in the colloquial sense of the word or a mere corrective addition. In other words, the mashup of Levinas and Heidegger, as is the case with all interesting digital recombinations, does not try to define the one in terms of the other (or vice versa) but seeks to preserve their specific distance and unique differences in an effort to articulate what the other had forgotten, left out, or passed over. "As supplementing each other," Benso concludes, "Levinas and Heidegger remain external, exterior, other, each not defined as the other than the same. But still as supplements, each of them offers the other that remainder that the other leaves unthought" (ibid., 129). The mashup therefore exceeds the controlling authority and comprehension of either thinker, producing an unauthorized hybrid that is neither one or the other nor a synthetic combination of the two that would sublate difference in a kind of Hegelian dialectical resolution.

According to Derrida, two different meanings cohabit, oddly although necessarily, in the notion of the supplement. The supplement is a surplus, an addition, a fullness that enriches another fullness. Yet the supplement is not only an excess. A supplement supplements. Its addition aims at replacement. It is as if it filled a void, an anterior default of a presence. It is compensatory and vicarious, "its place is assigned in the structure by the mark of an emptiness" (Derrida 1976, 144–145). Neither Heidegger nor Levinas need each other. Yet, in both there is a remainder that is not described, that is forgotten in the meditation. In Heidegger it is ethics, in Levinas it is things. (Ibid.)

For Heidegger, as Benso correctly points out, the thing was and remained a central issue throughout his philosophical project: "The question of things, Heidegger remarks at the beginning of a 1935/36 lecture course later published as *Die Frage nach dem Ding*, is one of the most ancient, venerable, and fundamental problems of metaphysics" (ibid., 59). Benso's attentive reading demonstrates how this "question concerning the thing" had been operationalized in Heidegger's own work, beginning with *Being and Time*'s critical reconsideration of Husserl's phenomenology, which had directed analytical efforts "to the things themselves" (Heidegger 1962, 50), and proceeding through numerous lecture-courses and publications that address *What Is a Thing?* (Heidegger 1967), "The Thing" (Heidegger 1971b), and "The Origin of the Work of Art" (Heidegger 1977a). And she charts Heidegger's thought about the thing by identifying three distinct phases:

the instrumental mode of disclosure, the artistic mode of disclosure, and the possibility of an ethical mode of disclosure.

The Instrumental Mode of Disclosure Heidegger's initial thinking of things is developed in his first and probably best-known work, *Being and Time*. In this early work, all things are accommodated to and comprehended by what Heidegger calls *das Zeug* or equipment (Heidegger 1962, 97). "The Greeks," Heidegger writes in his iconic approach to the subject, "had an appropriate term for 'Things': πράγματα—that is to say, that which one has to do with in one's concernful dealings (πρᾶξις). But ontologically, the specific 'pragmatic' character of the πράγματα is just what the Greeks left in obscurity; they thought of these 'proximally' as 'mere Things.' We shall call those entities which we encounter in concern *equipment* [*Zeugen*]. In our dealings we come across equipment for writing, sewing, working, transportation, measurement" (ibid., 96–97). According to Heidegger, the ontological status or the kind of being that belongs to such equipment is primarily disclosed as "ready-to-hand" or *Zuhandenheit*, meaning that something becomes what it is or acquires its properly "thingly character" in coming to be put to work by human *Dasein* for some particular purpose (ibid., 98). A hammer, one of Heidegger's principal examples, is for hammering; a pen is for writing; a shoe is for wearing. Everything is what it is in having a *for which* or a destination to which it is to be put to use.

"This does not necessarily mean," Benso (2000, 79) explains, "that all things are tools, instruments which *Dasein* effectively uses and exploits, but rather that they disclose themselves to *Dasein* as endowed with some form of significance for its own existence and tasks." Consequently, for Heidegger's analysis in *Being and Time*, the term *equipment* covers not only the disclosure of artifacts—and Heidegger provides a litany of such things, shoes, clocks, hammers, pens, tongs, needles—but the things of nature as well. Understood in this fashion, natural things either are encountered in the mode of raw materials—"Hammer, tongs, and needle refer in themselves to steel, iron, metal, mineral, wood, in that they consist of these" (Heidegger 1962, 100). "Or nature can be," as Benso (2000, 81) points out, "encountered as the environment in which *Dasein* as *Geworfen* exists. Again, the ecological conception of nature is disclosed with reference to and through its usability, so that the forest is always the wood usable for timber, the mountain a quarry of rock, the river is for producing water-power, the wind is 'wind in the sails.'" Everything is what it is and has its own unique being only insofar as it is always and already accommodated to and comprehended by human *Dasein*'s own concernful dealings.

In this mode of *primordial disclosure*, however, the things as such are virtually transparent, unnoticeable, and taken for granted. In being considered for something else, that which is ready-to-hand immediately and necessarily recedes from view and is manifest only insofar as it is useful for achieving some particular purpose—only to the extent that it is "handy." For this reason, the equipmentality of things as such only obtrudes and becomes conspicuous when the equipment fails, breaks down, or interrupts the smooth functioning of that which is or had been at hand. "The equipmental character of things is explicitly apprehended," Benso (2000, 82) writes, "*via negativa* when a thing reveals its unusability, or is missing, or 'stands in the way' of *Dasein*'s concern." In these circumstances, the thing comes to be disclosed as "presence-at-hand" or *Vorhandenheit*. But, and this is an important qualification, presence-at-hand is, strictly speaking, a derived, deficient, and negative mode of disclosure. What is merely present-at-hand comes forth and shows itself as such only when some thing has become conspicuously *un-ready-to-hand* (Heidegger 1962, 103). For the Heidegger of *Being and Time*, therefore, everything—whether an artifact, like a hammer, or one of the many things of nature, like a forest— is primordially disclosed as something to be employed by and for human *Dasein*. And anything that is not ready-to-hand is merely present-at-hand but only as something derived from and no longer ready-to-hand. "In *Being and Time*," Benso (2000, 94) explains, "all things, that is, all non-*Dasein* entities, were assimilated as *Zeug* or as modalities thereof, and even natural entities were equipment-to-be."

Heidegger therefore effectively turns everything—whether technological artifact or natural entity—into an instrument that is originally placed in service to and disclosed by human *Dasein* and its own concernful dealings with the world. In this early text, then, Heidegger does not think technology as a thing but, through his own mode of thinking about things, turns everything into something technological—that is, an instrument placed in service to and primarily disclosed by human *Dasein*'s own interests and concerns. In fact, Heidegger will later explicitly connect the dots, specifying *Zeug* as something properly belonging to the technological, in the essay "The Question Concerning Technology": "The manufacture and utilization of equipment [*Zeug*], tools, and machines, the manufactured and used things themselves, and the needs and ends that they serve, all belong to what technology is" (Heidegger 1977a, 4–5). Consequently, despite the promise to consider things otherwise, to make good on the promise of phenomenology's "return to things themselves," *Being and Time* accommodates all things to the conventional anthropocentric and instrumentalist view that Heidegger himself will criticize in his later work. "The much advocated 'return to things,'" Benso concludes, "remains only partially a return *to things* as such. 'The thingness must be something unconditioned,' Heidegger claims. What Heidegger thematizes in *Being and Time* and related works, however, is that *Dasein* is at the origin of the way of being of things. That is, the thingness of things is not truly unconditioned, as his inquiry into the question of things had promised" (Benso 2000, 92).

The Artistic Mode of Disclosure   Because of this rather conventional (albeit one that is carefully and systemically articulated) outcome, "the question of things needs to be asked again" (ibid.), and Benso calls the next phase of involvement the "artistic modes of disclosure." This phase of the thinking of things is situated in "The Origin of the Work of Art," which Heidegger first delivered as a public lecture in 1935. Although the objective of this text, as immediately communicated by its title, was to ascertain the "origin of the work of art," this goal could only be achieved, Heidegger (1971a, 20) argues, by "first bring to view the thingly element of the work." "To this end," he writes, "it is necessary that we should know with sufficient clarity what a thing is. Only then can we say whether the art work is a thing" (ibid.). In response to this charge, Heidegger begins the investigation of the origin of the work of art by turning his attention to things. Although he recognizes that the word "thing," in its most general sense, "designates whatever is not nothing," he also notes that this characterization is "of no use to us, at least immediately, in our attempt to delimit entities that have the mode of being of a thing, as against those having the mode of being of a work" (ibid., 21).

In order to provide a more attentive consideration of the matter, Heidegger develops what Benso calls "a tripartition." In the first place (and "first" is formulated and understood both in terms of expository sequence and ontological priority), there are "mere things," which, on Heidegger's account, designate "the lifeless beings of nature" (e.g., a stone, a clod of earth, a piece of wood, a block of granite). Mere things, Heidegger argues, are not directly accessible as such. They always and already withdraw and hold themselves back. If we were to impose Levinasian terminology on this, one might say that "mere things" constitute the extreme limit of alterity and exteriority. This is, according to another essay by Heidegger from the same period, namely *What Is a Thing?*, a kind of Kantian "thing-in-itself," but without the metaphysical baggage that this concept entails. Because of this, it is only possible, as Benso (2000, 99) describes it, "to approach the thing-being of things not in mere things but in pieces of equipment, despite the fact that the purity of their thing-being has been lost in favor of their usability."

Consequently, and in the second place, there are the "objects of use," the utensils or equipment (*Zeugen*) that had been analyzed in *Being and Time*, including such sophisticated instruments as "airplanes and radio sets" as well as "a hammer, or a shoe, or an ax, or a clock" (Heidegger 1971a, 21). But "to achieve a relation with things that lets them be in their essence, a thematic suspension of the usability of equipment is required" (Benso 2000, 101). In *Being and Time* this "thematic suspension" had occurred in the breakdown of equipment that revealed the mere present-at-hand. This is no longer considered sufficient. "It remains doubtful," Heidegger (1971a, 30) writes in something of a revision of his earlier efforts, "whether the thingly character comes to view at all in the process of stripping off everything equipmental." One discovers things, a pair of shoes for example, "not by a description and explanation of a pair of shoes actually present; not by a report about the process of making shoes; and also not by the observation of the actual use of shoes occurring here and there" (ibid., 35). This important work is performed by the work of art, specifically for Heidegger's analysis a painting by Van Gogh. "The art work let us know what shoes are in truth" (ibid.). It is in the painting of the shoes, Heidegger contends, that this particular being, "a pair of peasant shoes, comes in the work to stand in the light of its being" (ibid., 36). Consequently, as Benso (2000, 101) points out, a shift occurs in the place and work of disclosure: "The work of art thus substitutes for *Dasein*, thanks to whose constitution as being-in-the world, in *Being and Time*, the disclosure occurred." Or as Benso's chapter titles indicate, there is a move from the "instrumental modes of disclosure" to "artistic modes of disclosure."

According to Benso's evaluation, "the move to artistic disclosure is certainly beneficial to the thing-being of things, if compared to the earlier instrumental account provided in *Being and Time*" (ibid., 102). This is because artistic disclosure is, on her reading, considerably less violent and more respectful of things. In particular, the artistic mode of disclosure disputes and even undermines the anthropocentric privilege and its instrumentalist notion of things. "What is abandoned through artistic disclosure," Benso writes, "is *Dasein*'s instrumental attitude" (ibid., 103). Artistic disclosure, therefore, shifts the place of the disclosure of things from the concernful dealings of human *Dasein* to the work of art and in doing so deflects the anthropological and instrumentalist approach to things. This means that "works of art, although humanly made, are self-sufficient" (ibid., 104). Or as Heidegger (1971a, 40) explains, "it is precisely in great art . . . that the artist remains inconsequential as compared with the work, almost like a passageway that destroys itself in the creative process for the work to emerge."

But artistic disclosure, although an obvious advance over the instrumental modes of disclosure introduced in *Being and Time*, is not without problems. It is, as Benso (2000, 104–105) argues, "not completely innocent in its exposure of the thingly character of things." On the one hand, Heidegger's articulation of artistic disclosure is exclusive, if not snobbish. "It is only great art," Benso points out, "that is capable of bringing forth a happening of the truth" (ibid., 105). Consequently, all kinds of things may get lost, that is, not get disclosed, in our everyday involvements that are neither artistic nor great. "Some essential part of the thing-being of things," Benso worries, "will be irreversibly lost in the process of everyday concern with things" (ibid., 107).

On the other hand, despite Heidegger's claims to the contrary, a kind of violence remains at work in the work of art. "The violence," Benso argues, "lies in the fact that the disclosure originates not from within the thing itself, in its earthy, dark, reticent, obscure depth, but from outside the thing, from art" (ibid., 108). And this violence is, according to Benso's reading, a technical matter: "That the potential for violation is inherent in the artistic activity had been understood clearly by the Greeks, which used the same word, *techne* [τέχνη], to express both art and that kind of knowledge which will later give birth to technology and its aberrations. Art is not the self-disclosure of things; rather, it is an external act of disclosing" (ibid.). This is, for Benso at least, not adequate. Things should, in her opinion, disclose their being from themselves. In other words, the disclosure "should originate from what is already earth—namely, from the thing-being of things" (ibid.). The work of art, therefore, is already an aberrant distraction and a form of technical mediation that one should be able to do without. "The recourse to art seems to be an unnecessary superimposition in which art acts as a mediator between things and the disclosure of their thing-being. As Levinas has taught his readers, though, like all mediations such an interference endangers the risk of several violations. Among them, misunderstanding, neglect, abuse, betrayal of the thing-being of things" (ibid.).

An Ethical Disclosure? Although constituting a clear advance over the instrumental mode, the artistic mode of disclosure is, Benso contends, still too violent, still too *technological* to be respectfully attentive to the thing-being of things. It is, therefore, only in the later works, especially the essay "The Thing," that Heidegger, according to Benso's interpretation, finally gets it right. "In 'The Thing,'" Benso explains, "Heidegger chooses a rather obvious case of things, an artifact—a jug, which in *Being and Time* would have been considered, at best, a piece of equipment, and whose truth, according to 'The Origin of the Work of Art,' would have been disclosed only through its appearance in/as a work of art" (ibid., 112). In "The Thing," things are entirely otherwise. "The jug," Heidegger (1971b, 177) writes, trying to gathering up all the threads of his analysis, "is a thing neither in the sense of the Roman *res*, nor in the sense of the medieval *ens*, let alone in the modern sense of object. The jug is a thing insofar as it things. The presence of something present such as the jug comes into its own, appropriatively manifests and determines itself, only from the thinging of the thing." This third mode of disclosure, which is a disclosure of things from things, requires a radically different kind of response. It necessitates, as Benso interprets it, a mode of responding to the alterity of things that does not take the form of a violent imposition of external disclosure but which can let beings be—what Heidegger calls *Gelassenheit*. As Benso (2000, 123) explains: "Neither indifference nor neglect, neither laxity nor permissiveness, but rather relinquishment of the metaphysical will to power, and therefore acting 'which is yet no activity,' *Gelassenheit* means to abandon oneself to things, to let things be."

Whether this is an accurate reading of Heidegger or not, it produces something that is, for Benso at least, sufficient. It is in these later essays, Benso concludes, that "the thingness of things has been finally achieved by Heidegger, and has been achieved in its fundamental character of alterity: unconditioned alterity, because things are unconditioned; absolute alterity, because the alterity of things does not stem from an oppositional confrontation with mortals, or divinities, which are rather appropriated by and relocated with the alterity of things" (ibid., 119). And because of this, it is here that Benso finally discovers, in Heidegger's own text, the possibility of asking about what she calls an "ethical disclosure." "What kind of relation is this relation, which, by extension, encompasses also the relation between *Gelassenheit* and things? Although Heidegger's work is silent, this relation could modestly be called ethics, if, as in Levinas, ethics is understood as the place of love for what remains and insists on remaining other. Things thus impose an imperative which comes close to an ethical demand. They request an act of love—ethics—which lets things be as things. . . . Heidegger, however, will never explicitly thematize the ethical character of such an act" (ibid., 123).

In identifying this possibility of an ethics in Heidegger's thinking of things, Benso likewise identifies the point of contact with Levinas, who, although not thinking the alterity of things, provides articulation of the kind of ethics that Heidegger had left unthematized. Consequently, the mashup of these two thinkers, like any well-devised and executed remix, is not something that is forced on them from the outside but seems to show itself by way of careful attention to the original texts that come to be combined. And what this mashup produces is something unheard of—a mode of thinking otherwise about things and ethics that could be called "an ethics of things":

The expression "ethics *of* things," as the result of the supplementarity of Levinas and Heidegger, acquires a double meaning: it is *of* things, as the place where things can manifest themselves in their reality as the guardians and receptacle of the Fourfold, and from their receptivity can appeal to humans to dwell by them. But it is *of* things also in the sense that humans are compelled by things to respond to the demands placed upon them and shape their behavior in accordance to the inner mirroring of things. Things signify both a subject and object for ethics. Of things means thus the directionality of a double movement: that which moves out from the things to reach the I and the other, and that which, in response to the first moves from the I and the other to reach the things and to be concerned by them. The first movement is that of the demand or the appeal that things place on human beings by their mere impenetrable presencing there. It is the thingly side of the ethics of things. The second movement is that of tenderness, as the response to the demand and the properly human configuration of the ethics of things. (Ibid., 142)

As innovative as this proposal sounds, especially for the encounter with those things that call to us in their "obsessive appeal" to be recognized as other and to which we are called to provide some kind of response, one thing continually escapes Benso's "ethics of things," and that is the machine.

Although Heidegger began his thinking of things by appropriating everything into an instrument of human *Dasein*'s concernful dealings with the world, the later essays, the essays that Benso argues provides a more adequate thinking of the thing, differentiate things from mere objects of technology. In "The Question Concerning Technology," an essay that dates from the same period as "The Thing," Heidegger distinguishes "modern technology" as a totalizing and exclusive mode of revealing that threatens the disclosure of things by converting them into resources or what Heidegger calls *Bestand* (standing-reserve). "The coming to presence of technology threatens revealing, threatens it with the possibility that all revealing will be consumed in ordering and that everything will present itself only in the unconcealedness of standing-reserve" (Heidegger 1977a, 33). According to Benso's (2000, 144) reading, this means that technology does not let things be but "diminishes things to objects of manipulation" through a "perverted logic of powerfulness" that leaves things "deprived of their thingly nature." Or to look at it from the other side, "the appeal things send out is an invitation to put down the dominating modes of thought, to release oneself to things, and only thus to let them be as things, rather than as objects of technical and intellectual manipulation" (ibid., 155).

This reading of Heidegger, which it should be noted is not entirely accurate insofar as it passes over Heidegger's (1977a, 28) remark that "the essence of technology must harbor in itself the growth of the saving power," leads Benso to make romanticized claims about pre–industrial era peasants and their imagined direct and immediate connection to the things of the earth. "Contrary to Heidegger's claim in 'The Origin of the Work of Art,' but in accordance with the future line of development of his meditation on things, a more respectful relation with things, dictated from the very thing-being of things, seems to take place not in Van Gogh's artistic painting of the shoes, but rather in the attitude of the peasant woman—not in the usability to which she submits her shoes, but in her depending on the shoes' reliability" (Benso 2000, 109). In other words, contrary to Heidegger's privileging of art, artists, and poets in his later writings, it is the peasant who "may be close to the thing-being of things, respectful and preservative of their earthy character" (ibid.). Benso's analysis, therefore, privileges a supposed preindustrial and romantic notion of the European peasant (which is, in fact, a modern and industrial fabrication projected backward onto a past that never actually existed as such) and her assumed direct and immediate involvement with real things.

The machine, as a mere thing, already has no place in the nontraditional ethics of Levinas. It is, as far as Levinas is concerned, and even those individuals like Calarco who work to radicalize his thinking, always and already otherwise than the Other. A machine might have a well-designed and useful *interface*, but it does not and will never have a *face*. Likewise, machines do not belong to or have an appropriate place within Heidegger's conceptualization of things. Although Heidegger endeavored to address things and in particular sought to accommodate thinking to address the thing-being of things, the machine, as a technological object, remains otherwise than a thing. It is an object, which, as far as Benso is concerned, is neither a thing nor the opposite of a thing, that is, nothing. For this reason, Benso's mashup of these two thinkers, despite its promise and her careful attention to the material at hand, is not able to accommodate or address itself to the machine question. And this is mainly because of Benso's initial decision concerning the terms and conditions of the meeting. She endorses what she calls "the narrow path of affirmation"—things and ethics—and immediately excludes the negative mode of the encounter— nonthings and non-ethics. The machine does not get addressed or have a place in the course of this encounter, because it is always and already on the excluded side of things—the side of nonthings and non-ethics. In addressing herself to the affirmative mode of the Heideggerian and Levinasian mashup, Benso decides, whether intended or not, to leave the machine out of the (re)mix. It is, therefore, from the very beginning situated outside and beyond the space of consideration, in excess of the Levinasian–Heideggerian mashup, and remains both anterior and exterior to this attempt to think things and ethics otherwise.

The machine does not have a place in *The Face of Things* (and it should be noted that the term "machine" does not appear anywhere within the text), because Benso's approach had effectively excluded it prior to the investigation. This outcome, it should be noted, is not necessarily some deliberate failing of or fault attributable to this particular writer or her writing. In the process of composing her text and its arguments, a decision had to be made; a cut needed to occur and could not have been avoided. That is, the analysis could not have proceeded, could not have even gotten underway, without instituting some exclusive decision about the terms and conditions of the encounter. These decisions, whether made in the course of a philosophical mashup between Levinas and Heidegger or in the process of remixing, for instance, Madonna and the Sex Pistols (Vidler 2007), is always strategic, calculated, and undertaken for a particular purpose, at a particular time, and in view of a specific objective. The problem—the persistent and seemingly inescapable problem—is that when the cut is made, something always and inevitably gets left out and excluded. And this something, which is, strictly speaking neither some-thing nor no-thing, is, more often than not, the machine. For this reason, it can be said that the machine has been always and already the other of the other, no matter how differently alterity comes to be rethought, remixed, or radicalized.

3.4.3 Machinic Others
Although Calarco's and Benso's texts have nothing explicit to say with regard to the machine, others have endeavored to provide a thinking that explicitly addresses such things. Lucas Introna, like Benso, endeavors to articulate an ethics of things, or, more precisely described, "the possibility of an ethical encounter with things" (Introna 2009, 399). Whereas Benso produces a mashup of Heideggerian and Levinasian thought by way of a careful engagement with their texts, Introna concentrates his efforts on Heidegger and does so mainly by way of secondary literature, especially Graham Harman's (2002) reading of "tool-being" in *Being and Time*. According to Introna, Harman advances an interpretation of the Heideggerian text that runs counter to the orthodox reading in at least two ways. "He argues that ready-to-handness (*Zuhandenheit*) already 'refers to objects insofar as they withdraw from human view into a dark subterranean reality that never becomes present to practical action' (Harman 2002, 1). He further argues, rather controversially, that *Zuhandenheit* is not a modification, or mode of revealing reality, which is uniquely connected to the human *Dasein*. Rather, *Zuhandensein* is the action of all beings themselves, their own self-unfolding of being" (Introna 2009, 406). In this way, then, Harman (2002, 1) advances what he calls an "object-oriented philosophy" that effectively reinterprets Heidegger as Kant by distinguishing between

"objects as explicitly encountered (*Vorhandenheit*) and these same objects in their withdrawn executant being (*Zuhandenheit*)" (Harman 2002, 160).

The result of following this unconventional and rather conservative interpretation of Heidegger (which becomes apparent when it is compared to the analysis of the same material provided in Benso's work) is something Introna (2009, 410) calls "the ethos of *Gelassenheit*." Introna characterizes *Gelassenheit*, which is one of the watchwords of the later Heidegger, as a mode of comportment that gives up "that representational and calculative thinking (or comportment) by which human beings dispose of things as this or that being" (ibid.) and "lets things be, as they are, in their own terms" (ibid., 409–410). This effort is, on the one hand, something of an advance over Benso's innovation insofar as Introna, following Harman's interest in equipment and tools, does not restrict things to natural objects but specifically addresses the relationship with or comportment toward technological artifacts like cars, pens, and chairs (ibid., 411). On the other hand, however, Introna's "ethos of dwelling with things" is far less successful than what Benso is able to achieve by way of her mashup. In effect, what Introna offers under the banner "ethos of *Gelassenheit*" is really little more than a sophisticated version of the lightbulb joke concerning German engineering: "Q: How many German engineers does it take to change a light bulb? A: None. If it is designed correctly and you take care of it, it should last a lifetime." In the end, what Introna provides, mainly because of his reliance on Harman's limited "object oriented" interpretation of the early Heidegger, is little more than another articulation of an ontocentric ethical comportment that looks substantially similar to Floridi's IE. What Introna calls "the impossible possibility of a very otherwise way of being with things" turns out to be just more of the same.

Coming at things from the other side, Richard Cohen takes up and directly examines the point of contact between ethics, especially in its Levinasian form, and what he calls "cybernetics."[3] The immediate target of and point of departure for this critical investigation is Sherry Turkle's *Life on the Screen* (1995) and Introna's 2001 essay "Virtuality and Morality: On (Not) Being Disturbed by the Other," which Cohen situates on opposite sides of a debate:

Thus Turkle celebrates cybernetics' ability to support a new form of selfhood, the decentered and multiple self (or rather selves). The multiple self cannot be held accountable in the same way that the integral self of morality is held accountable. Cybernetics, then, liberates the traditional self for the freedom of multiple selves.

Introna, for his part, seems to be arguing the reverse point when he condemns information technology. But in fact he too credits cybernetics with a radical transformation, or the possibility of a radical transformation of morality. Because it mediates the face-to-face relation that, according to Levinas's ethical philosophy, is the very source of morality, cybernetics would be the destruction of morality. (Cohen 2010, 153)

What Cohen argues, then, is that Turkle and Introna, although seemingly situated on the opposite sides of a debate—what the one celebrates, the other reviles—actually "adhere to the same metainterpretation of cybernetics whereby it is considered capable of radically transforming the human condition" (ibid.).

Cohen, who does not take sides in this debate, takes aim at and contests this common assumption, and his argument follows, with little or no critical hesitation, the anthropological and instrumentalist position that had been identified by Heidegger: "The question, then, is whether computer technology produces a radical transformation of humanity, or whether, in contrast, it is simply a very advanced instrument, tool, or means of information and image processing and communication that is itself morally neutral" (Cohen 2010, 153). According to Cohen's reading, Turkle and Introna occupy and defend the former position, while he endeavors to take up and champion the latter. In making this argument, Cohen not only reasserts and reaffirms the standard instrumentalist presumption, asserting that the "so-called computer revolution" is "not as radical, important, or transformative as many of its proponents, such as Turkle, or its detractors, such as Introna, would have it" (ibid., 154), but also interprets Levinas as both endorsing and providing support for this traditional understanding of technology.

This is, of course, a projection or interpretation of Levinasian philosophy. Levinas himself actually wrote little or nothing about the machine, especially computers and information technology. "Admittedly," Cohen points out, "although he developed his philosophy over the second half of the twentieth century, Levinas did not specifically write about cybernetics, computers, and information technology" (ibid.). Even though Levinas lived and worked at a time when computers were becoming widely available and telecommunications and networking technology had proliferated at a rate of acceleration that is commonly understood to be unprecedented in human history, Levinas, unlike Heidegger, never took up and engaged the question concerning technology in any direct and explicit way. Cohen not only interprets this silence as supporting the anthropological and instrumentalist position but makes the further argument that Levinas's ethics, despite its lack of providing any explicit statement on the subject, "is ideally suited to raise and resolve the question of the ethical status of information technology" (ibid.).

Cohen, therefore, takes the Levinasian text at its word. The face-to-face encounter that constitutes the ethical relationship is exclusively human, and as such it necessarily marginalizes other kinds of others, specifically the old Cartesian couple, animals and machines. This exclusivity is not, Cohen believes, immoral or ethically suspect, because the machine and the animal do not, for now at least, constitute an Other:

Oddly enough computers do not think—are not human—not because they lack human bodies, but because like stones and animals they lack morality. They are indeed embodied, but their embodiment, unlike human embodiment, is not constituted—or "elected"—by an ethical sensitivity. Computers, in a word, are by themselves incapable of putting themselves into one another's shoes, incapable of intersubjective substitution, of the caring for one another which is at the core of ethics, and as such at the root of the very humanity of the human. (Ibid., 163)

For this reason, machines are instruments that may come to be interposed between the self and the Other, mediating the face-to-face encounter, but they remain mere instruments of human interaction. "Cybernetics thus represents a quantitative development: increases in the speed, complexity, and anonymity of communications already inherent in the printing press, an increase in the distance—but not a radical break—from the immediacy, specificity, singularity, and proximity of face-to-face encounters" (ibid., 160). To put it another way, computers do not have a face and therefore do not and cannot participate in the face-to-face encounter that is the ethical relationship. Instead, what they offer is an interface, a more or less transparent medium interposed and standing in between the face-to-face encounter. In making this argument, then, Cohen effectively repurposes Levinasian ethics as media theory.

The main problem for Cohen's argument is that he misunderstands both terms that are conjoined in the title of his essay. On the one hand, he misunderstands or at least significantly misrepresents cybernetics. According to Cohen's analysis, cybernetics is just "the most recent dramatic development in the long history of communications technology" (ibid., 159). Consequently, he understands and utilizes the word "cybernetics" as an umbrella term not only for information technology and computers in general but also for the specific applications of e-mail, word processing, and image manipulation software. This is not only terribly inaccurate but also unfortunate.

First, cybernetics is not a technology, nor is it a conglomerate of different information and communication technologies. It is, as originally introduced and formulated by its progenitor Norbert Wiener, the general science of communication and control. "We have decided," Wiener writes in *Cybernetics*, a book initially published in 1948, "to call the entire field of control and communication theory, whether in the machine or the animal, by the name *Cybernetics*, which we form from the Greek χυβερνήτης or *steersman*" (Wiener 1996, 11). Cybernetics, then, is not a kind of technology or a particular mode of technological application but a theory of communication and control that covers everything from individual organisms and mechanisms to complex social interactions, organizations, and systems.[4] According to Carey Wolfe, cybernetics introduced a radically new way of conceptualizing and organizing things. It proposed, he argues, "a new theoretical model for biological, mechanical, and communicational processes that removed the human and *Homo sapiens* from any particular privileged position in relation to matters of meaning, information, and cognition" (Wolfe 2010, xii). Cohen, therefore, uses the word "cybernetics" in a way that is neither informed by nor attentive to the rich history of the concept. And in the process, he misses how cybernetics is itself a radical, posthuman theory that deposes anthropocentrism and opens up thoughtful consideration to previously excluded others. Because of this, the editors of the journal in which Cohen initially published his essay provide the following explanatory footnote as a kind of excuse for this misinterpretation: "Richard Cohen uses the word 'cybernetics' to refer to all forms of information and communication technology" (Cohen 2000, 27).

But, and this is the second point, Cohen does not just "get it wrong" by misrepresenting the concept or misusing the word "cybernetics," which could, in the final analysis, always be excused and written off as nothing more than a mere terminological misstep. Rather, by doing so, Cohen actually "defeats his own purpose," to use a phrase popularized by Robert De Niro's Jake LaMotta in Martin Scorsese's *Raging Bull* (1980). In particular,

Cohen, by misconstruing what cybernetics entails, misses the more fundamental and potent point of contact between it and Levinas's philosophy. If Levinasian ethics is, as Cohen presents it, based on an intersubjective, communicative experience of or encounter with the Other, then cybernetics as a general theory of communication not only addresses itself to a similar set of problems and opportunities but, insofar as it complicates the anthropocentric privilege and opens communication with and to previously excluded others, also provides an opportunity to "radicalize" Levinasian thought by asking about other forms of otherness. Consequently, cybernetics may be another way to articulate and address the "otherwise than being" that is of central concern in Levinasian ethics. And it is, we should remember, Heidegger who had prepared the ground for this possibility, when, in the course of his 1966 interview published in the German magazine *Der Spiegel*, he suggested that what had been called philosophy was in the process of being replaced by cybernetics.

Derrida famously picks up this thread in *Of Grammatology*, demonstrating how cybernetics, even as it strains against the limits of metaphysics, is still circumscribed by a certain concept of writing: "And, finally whether it has essential limits or not, the entire field covered by the cybernetic *program* will be the field of writing. If the theory of cybernetics is by itself to oust all metaphysical concepts—including the concepts of soul, of life, of value, of choice, of memory—which until recently served to separate the machine from man, it must conserve the notion of writing, trace, gramme, or grapheme, until its own historico-metaphysical character is also exposed" (Derrida 1976, 9). Cohen, therefore, fabricates a derived caricature of cybernetics—one that turns it into a mere technological instrument so that it can be manipulated as a tool serving Cohen's own argument, which reasserts the instrumentalist understanding of technology. In doing so, however, Cohen not only risks getting it wrong but, more importantly, misses what he could have gotten right. In facilitating the conjunction of Levinasian ethics and cybernetics, Cohen introduces a potentially interesting and fruitful encounter between these two influential postwar innovations only to recoil from the radicality that this conjoining makes possible and to reinstitute what is perhaps the most reactionary and predictable of responses.

On the other hand, Cohen also risks misrepresenting ethics, and Levinasian ethics in particular. Although he recognizes and acknowledges "the

humanism of the other" (Levinas 2003) as it is construed in Levinas's philosophy, he does not, to his credit, take this to be an essential or even
absolute limit. It is possible that things could, at some point in the future,
be otherwise. "I have mentioned," Cohen (2010, 165) admits, "the possibility of animals and machines joining the one brotherhood of ethical sensitivity. In our day, however, moral responsibility and obligations have
their sources in human sensitivities, in the humanity of the human."
Cohen, therefore, appears to open up the boundaries of Levinasian philosophy to the possibility of addressing another kind of otherness. In other
words, even though the Other has been and remains exclusively human,
it may be possible, Cohen suggests, that an animal or a machine might,
at some point in the future, become capable of gaining access to the fraternal "brotherhood of ethical sensitivity." For now, however, animals and
machines, the old Cartesian couple, remain, at least as Cohen sees it, exterior to Levinas's reconsideration of exteriority.[5] Or to put it another way,
the animal-machine remains, at least for the time being, the other of Levinas's Other. This conclusion requires at least two comments.

First, Cohen, to his credit, does not simply pass over or remain silent
about the possibility of repurposing Levinas's philosophy so as to be able
to address itself to others—especially those other forms of otherness found
in the animal and machine. But he unfortunately at the same time ends
up confirming the Cartesian decision, postponing the moral challenge of
these others by deferring it to some future time. To make matters even
more complicated, even if and when, at this future moment, we succeed
in creating sentient machines like "the policeman-robot in the movie
Robocop or the character called Data in the television series *Star Trek: The
Next Generation*" (ibid., 167), they will, Cohen believes, still be subordi-
nated and considered subservient to the moral center exclusively constituted and occupied by the human organism. "One has," Cohen writes in
a footnote addressing these two science fiction characters, "moral obligations and responsibilities first to organisms, indeed to human organisms,
before one has moral obligations and responsibilities to machines that
serve humans or other organisms. . . . Note: to give priority to moral obligations and responsibilities to humans is not to deny the bearing of moral
obligations and responsibilities toward the nonhuman, whether organic
or inorganic. It is rather to locate the true source of moral obligations
and responsibilities" (ibid.). Although recognizing that other organic and

inorganic entities are not simply to be excluded form moral consideration *tout court,* Cohen still enforces the anthropocentric privilege, asserting that these others will remain, always and forever, subordinate to the human entity and his or her interests (and literally subordinate, in that they are only considered in the subordinate place of a footnote). In this way, then, Cohen both releases a possible challenge to the "humanism of the other" in Levinas's ethics and, at the same time, shuts it down by reinforcing and reasserting anthropocentric hegemony.

Second, although Cohen is open to the possibility that there may, at some point in the future, be other forms of otherness that would need to be taken into consideration, the way these others become Other is by achieving what Cohen (2010, 164) calls "the humanity of the human."

The humanity of the human does not arise from an animal or machine evidencing logic or the rationality of means and ends. Ants, termites, bees, and porpoises, after all, are rational in this sense. Rather, the humanity of the human arises when an animal or any being, is moved not by efficiency but by morality and justice. A being becomes moral and just when in its very sensibility, and across the pacific medium of language, it finds itself desiring an undesirable and insatiable service for the other, putting the other's need before its own. . . . If it happens that one day animals or machines become capable of independent moral sensitivity, then they too will enter into the unitary and unifying solidarity of moral agency. (Ibid., 164–165)

In other words, in order for these, for now at least, excluded others—namely, animals and machines—to be considered Other, that is to be admitted into "the unitary and unifying solidarity of moral agency," they will need to achieve that kind of "independent moral sensitivity" that is the very definition of the humanity of the human. They will, like Asimov's Andrew in the short story "Bicentennial Man," need to become not just rational beings but human beings.

This is, whether it is ever explicitly identified as such or not, a radical form of anthropocentrism, one that is much more exclusive and restrictive than what has been advanced by others under the umbrella term "person-ism." In this way, then, Cohen not only reasserts that ancient doctrine whereby "man is the measure of all things" but seems to advance a posi-tion that would, structurally at least, be contrary to both the letter and spirit of Levinas's own moral innovations. That is, the decision Cohen institutes concerning these other forms of otherness seems, despite what he says in support of Levinasian ethics, to enact a distinctly anti-Levinasian gesture by reducing these others to the same. Animals and machines, as an other and like any other form of otherness, confront the self-assured enclosure of anthropocentric ethics. But rather than permitting this inter-ruption of the other to call into question this self-certainty and hegemony, Cohen imposes it on these others in that kind of violent gesture that Levinas had sought to criticize. In this way, then, Cohen's argument appears to be exposed to the charge of what Jürgen Habermas (1999, 80), following Karl-Otto Apel, has called a "performative contradiction," whereby what is explicitly stated and endorsed is called into question and undermined by the way it is stated and endorsed. This is, it should be pointed out, not necessarily some deficiency or inability that can or even should be attributed to the author of the text. It is rather an indication and evidence of the persistent and systemic difficulty inherent in address-ing others, especially the animal and its other, the machine.

3.5 Ulterior Morals

"Every philosophy," Benso (2000, 136) writes in a comprehensive gesture that performs precisely what it seeks to address, "is a quest for wholeness." This objective, she argues, has been typically targeted in one of two ways. "Traditional Western thought has pursued wholeness by means of reduc-tion, integration, systematization of all its parts. Totality has replaced wholeness, and the result is totalitarianism from which what is truly other escapes, revealing the deficiencies and fallacies of the attempted system" (ibid.). This is precisely the kind of violent philosophizing that Levinas identifies under the term "totality," and which includes, for him at least, the big landmark figures like Plato, Kant, and Heidegger. The alternative to this totalizing approach is a philosophy that is oriented otherwise, like that proposed and developed by Levinas and others. This other approach, however, "must do so by moving not from the same, but from the other, and not only the Other, but also the other of the Other, and, if that is the case, the other of the other of the Other. In this must, it must also be aware of the inescapable injustice embedded in any formulation of the other" (ibid.). What is interesting about these two strategies is not what makes them different from one another or how they articulate approaches that proceed from what appears to be opposite ends of the spectrum. What is interesting is what they agree on and hold in common in order to be situated as different from and in opposition to each other in the first place. Whether taking the form of autology or some kind of heterology, "they both share the same claim to inclusiveness" (ibid.), and that is the problem.

We therefore appear to be caught between a proverbial rock and a hard place. On the one hand, the same has never been inclusive enough. The machine in particular is from the very beginning situated outside ethics. It is, irrespective of the different philosophical perspectives that come to be mobilized, neither a legitimate moral agent nor patient. It has been and continues to be understood as nothing more than a technological instrument to be employed more or less effectively by human beings and, for this reason, is always and already located in excess of moral considerability or, to use that distinct Nietzschean (1966) characterization, "beyond good and evil." Technology, as Lyotard (1993, 44) reminds us, is only a matter of efficiency. Technical devices do not participate in the big questions of metaphysics, aesthetics, or ethics. They are nothing more than contrivances or extensions of human agency, used more or less responsibly by human agents with the outcome affecting other human patients. Consequently, technological artifacts like computers, robots, and other kinds of mechanisms do not, at least for the majority philosophical opinion, have an appropriate place within ethics. Although other kinds of previously excluded others have been slowly and not without considerable struggle granted membership in the community of moral subjects—women, people of color, some animals, and even the environment—the machine remains on the periphery. It exceeds and escapes even the best efforts at achieving greater inclusivity.

On the other hand, alternatives to this tradition have never quite been different enough. Although a concern with and for others promised to radicalize all areas of thought—identity politics, anthropology, psychology, sociology, metaphysics, and ethics—it has never been entirely adequate or suitably different. This is because such an effort has remained, if we might once again be permitted an allusion to Nietzsche (1986), "human, all too human." Many of the so-called alternatives, those philosophies that purport to be interested in and oriented otherwise, have typically excluded the machine from the space of difference, from the difference of difference, or from the otherness of the Other. Technological devices certainly have an interface, but they do not possess a face or confront the human user in a face-to-face encounter that would call for and would be called ethics.

This exclusivity is not simply "the last socially accepted prejudice" or what Singer (1989, 148) calls "the last remaining form of discrimination," which may be identified as such only from a perspective that is already open to the possibility of some future inclusion and accommodation. The marginalization of the machine is much more complete and comprehensive. In fact, the machine does not constitute just one more form of alterity that would be included at some future time. It comprises, as we have seen, the very mechanism of exclusion. "In the eyes of many philosophers," Dennett (1998, 233) writes, "the old question of whether determinism (or indeterminism) is incompatible with moral responsibility has been superseded by the hypothesis that mechanism may well be." Consequently, whenever a philosophy endeavors to make a decision, to demarcate and draw the line separating "us" from "them," or to differentiate who or what does and who or what does not have a face, it inevitably fabricates machines. The machine, therefore, exceeds difference, consisting in an extreme and exorbitant form of differentiation situated beyond and in excess of what is typically understood and comprehended as difference. It is otherwise than the Other and still other than every other Other. In other words, it remains excluded from and left out by well-intended attempts to think and address what has been excluded and left out. It is, to redeploy and reconfigure one of the titles to Levinas's books, otherwise than other or beyond difference.

The machine, therefore, constitutes a critical challenge that both questions the limits of and resists efforts at moral consideration, whether that takes the all-inclusive totalitarian form of the same or one or another of the alternative approaches that are concerned with difference. To put it another way, the machine occupies and persists in a kind of extreme exteriority that remains in excess of the conceptual oppositions that already organize and regulate the entire field of moral consideration—interior–exterior, same–different, self–other, agent–patient, subject–object, and so on. Asking the machine question, therefore, has a number of related consequences that affect not just where we go from here but also where we came from and how we initially got here.

First, there is a persistent and inescapable problem with words and terminology. Articulating the machine question and trying to address this form of extreme alterity that is otherwise than what is typically considered to be other, requires (as is clearly evident in this very statement) a strange contortion of language. This is not necessarily unique to the machine question; it is a perennial difficulty confronting any attempt "to think outside box" or in excess of what Thomas Kuhn had called "normal science." "Normal science," Kuhn (1996, 10) writes, "means research firmly based upon one or more past scientific achievements, achievements that some particular scientific community acknowledges for a time as supplying the foundation for its further practice." Normal science, therefore, establishes a framework, or paradigm, for investigation, a set of recognized procedures and methods for conducting research, and, perhaps most importantly, a shared vocabulary for asking questions and communicating results. Challenging this precedent and seeking to identify, name, or address "something" (which from the normalized perspective of the usual way of doing things would actually be considered "nothing") that has always and already been situated outside the scope of this conceptual field necessarily exceeds and resists the only language and concepts we have at our disposal. For this reason, there are typically two possible modes of responding to and articulating these paradigm shifting challenges—*paleonymy* and *neologism*.

Paleonymy is a Derridean (1981, 71) term fabricated from available Latin components to name the reuse and repurposing of "old words." Consequently, the word "paleonymy" is itself an instance of paleonymy. Using "an old name to launch a new concept" (ibid.) requires that the term be carefully selected and strategically reconfigured in order to articulate something other than what it was initially designed to convey. It therefore requires what Derrida characterizes as a double gesture: "We proceed: (1) to the extraction of a reduced predicative trait that is held in reserve, limited in a given conceptual structure, *named X*; (2) to the delimitation, the grafting and regulated extension of the extracted predicate, the name X being maintained as a kind of *lever of intervention*, in order to maintain a grasp on the previous organization, which is to be transformed effectively" (ibid.). This paleonymic operation is evident, for example, in Gilles Deleuze's *Difference and Repetition*, a 1968 publication that not only marked an important transition from Deleuze's earlier writings on the history of philosophy to the act of writing philosophy per se but also prefigured, as he suggests, the direction of all his subsequent publications, including those coauthored with Félix Guattari (Deleuze 1994, xv, xvii).

As is immediately evident from its title, *Difference and Repetition* is concerned with the "metaphysics of difference" and endeavors to formulate a different conceptualization of difference, that is, "a concept of difference without negation" (ibid., xx). "We propose," Deleuze writes in the text's preface, "to think difference in itself independently of the forms of representation which reduce it to the Same, and the relation of different to different independently of those forms which make them pass through the negative" (ibid., xix). *Difference and Repetition*, therefore, reuses and redeploys the old word *difference* in order to name a "new" and different concept of difference—one that cannot be reduced to negation and, as such, necessarily exceeds comprehension by the customary philosophical understanding of difference that had persisted from Plato to at least Hegel, if not beyond.

Neologism deploys a different but related strategy. "Neologism" is a rather old word, again comprising Latin roots, that identifies the fabrication of new words to name new concepts. Derrida's *différance*, for example, is a neologism for a nonconcept or quasi-concept that is, quite literally in this case, different from difference, or that marks a point of contact with and differentiation from the thinking of difference that had been situated in the history of philosophy. As Derrida (1981, 44) explains, "I have attempted to distinguish *différance* (whose *a* marks, among other things, its productive and conflictual characteristics) from Hegelian difference, and have done so precisely at the point at which Hegel, in the greater *Logic*, determines difference as contradiction only in order to resolve it, to interiorize it, to lift it up (according to the syllogistic process of speculative dialectics) into the self-presence of an ontotheological or onto-teleological synthesis." For Derrida, the visibly different *différance* indicates a different way to think and write of a difference that remains in excess of the Hegelian concept of difference. The machine question, therefore, challenges the available philosophemes, theoretical concepts, and extant terminology, necessitating linguistic contortions that seem, from the perspective of the normal way of doing things, curious and overly complicated. Whether one employs the strategy of paleonymy or neologism, articulating and addressing the machine question pushes language to its limits in an effort to force the available words to express that which remains in excess of what is considered possible or even appropriate.[6]

Second, and because of this, attempts to address what is and remains otherwise inevitably risk falling back into and becoming reappropriated by the established structures and protocols. Whether employing the strategy of paleonymy or neologism, efforts to think and write differently are always struggling against the gravitational force of existing structures, which understandably try to domesticate these extraordinary efforts and put them to work for the continued success of the established system of "normal science." This is what Žižek (2008b, vii), in an obvious but unacknowledged reworking of Kuhn, calls "Ptolemization." For this reason, any critical challenge to the status quo cannot be a "one off" or simply concluded or dispensed with once and for all. It is and must remain what Derrida (1981, 42) termed an "interminable analysis"—a kind of inexhaustible mode of questioning that continually submits its own achievements and advancements to additional questioning. Although Hegel (1969, 137) had called this kind of recursion[7] a "bad or spurious infinite" (*das Schlecht-Unendliche*), it is the necessary and inescapable condition of any and all critical endeavors.

For this reason, the machine question does not and cannot conclude with a definitive answer or even the pretense of supplying answers. The question, therefore, is not something to be resolved once and for all with some kind of conclusive and ultimate outcome. Instead, the result is a more sophisticated asking of the question itself. We began by questioning the place of the machine in ethics. It appeared, from the outset at least, to be a rather simple and direct inquiry. Either computers, AI's, robots and other mechanisms are a legitimate moral agent and/or patient, or they are not. That is, these increasingly autonomous machines either are responsible and accountable for what they decide and do, remain mere instruments in service to other interests and agents, or occupy some kind of hybrid in-between position that tolerates a mixture of both. Conversely, we either have a legitimate moral responsibility to these mechanized others, are free to use and exploit them as we desire without question or impunity, or cooperate with them in the formation of new distributed modes of moral subjectivity. In the course of pursuing this inquiry and following its various implications and consequences, however, all kinds of other things became questionable and problematic. In fact, it is in the face of the machine, if it is permissible to use this clearly Levinasian influenced turn of phrase, that the entire structure and operations of moral philosophy get put on the line. The machine question, therefore, is not some specific anomaly or recent crisis that has come into being alongside contemporary advancements in computers, artificial intelligence, robotics, artificial life, biotechnology, and the like. It is a fundamental philosophical question with consequences that reverberate down through the history of Western thought.

From one perspective, this outcome cannot help but be perceived as a rather inconclusive kind of ending, one that might not sit well with those who had anticipated and wanted answers or neatly packaged lists of dos and don'ts. In fact, this is precisely what is often expected of a work in ethics, especially applied ethics. And the expectation is not without a certain intuitive attraction: "Those of us who live and work in the 'real world' and need to make day-to-day decisions want to know what to do. What we want and what we need are answers to moral questions or if not answers, then at least guidelines to help us resolve these important questions." Instead of satisfying this expectation, things have ended otherwise. The investigation does not simply seek to answer whether and to what extent computers, robots, AI systems, and other mechanisms might be morally significant. Instead, or in addition, it releases a cascade of critical inquiry that intervenes in and asks about the very limits and possibilities of moral thinking itself. In this way, the machine is not necessarily a question *for* ethics; it is first and foremost a question *of* ethics.

Understood in this manner, the machine institutes a kind of fundamental and irresolvable questioning—one that problematizes the very foundation of ethics and causes us to ask about the ethicality of ethics at each stage of what appears to be a more inclusive approach to accommodating or addressing the differences of others. To put it another way, asking the machine question is not necessarily about getting it right once and for all. Rather, it is about questioning, again and again, what it is that we think we have gotten right and asking what getting it right has had to leave out, exclude, or marginalize in order to "get it right." To paraphrase Floridi (2008, 43), and to agree with his analysis in excess of the restricted interpretation he gives it, the machine question not only adds interesting new dimensions to old problems, but leads us to rethink, methodologically, the very grounds on which our ethical positions are based.

Finally, what this means for ethics is that Descartes, that figure who, at the beginning of the investigation, was situated in the role of the "bad guy," may have actually gotten it right despite himself and our usual (mis) interpretations of his work. In the *Discourse on the Method*, something of a philosophical autobiography, Descartes famously endeavored to tear down to its foundations every truth that he had come to accepted or had taken for granted. This approach, which in the *Meditations* comes to be called "the method of doubt," targets everything, including the accepted truths of ethics. With Descartes, then, one thing is certain: he did not want nor would he tolerate being duped. However, pursuing and maintaining this extreme form of critical inquiry that does not respect any preestablished boundaries has very real practical expenses and implications. For this reason, Descartes decides to adopt a "provisional moral code," something of a temporary but stable structure that would support and shelter him as he engaged in this thorough questioning of everything and anything:

> Now, before starting to rebuild your house, it is not enough simply to pull it down, to make provision for materials and architects (or else train yourself in architecture), and to have carefully drawn up the plans; you must also provide yourself with some other place where you can live comfortably while building is in progress. Likewise, lest I should remain indecisive in my actions while reason obliged me to be so in my judgments, and in order to live as happily as I could during this time, I formed for myself a provisional moral code consisting of just three or four maxims, which I should like to tell you about. (Descartes 1988, 31)

The four maxims include: (1) obeying the laws and customs of his country in order to live successfully alongside and with others; (2) being firm and decisive in action, following through on whatever opinion had come to be adopted in order to see where it leads; (3) seeking only to master himself instead of his fortunes or the order of the world; and (4) committing himself to the occupation of philosophy, cultivating reason and the search for truth (ibid., 31–33). Understood and formulated as "provisional," it might be assumed that this protocol would, at some future time, be replaced by something more certain and permanent. But Descartes, for whatever reason, never explicitly returns to the list in order to finalize things. This is, despite initial appearances, not a deficiency, failure, or oversight. It may, in fact, be the truth of the matter—that "all morality we adopt is provisory" (Žižek 2006a, 274), or, if you like, that ethics is provisional from the very beginning and all the way down. In this case, then, what would have customarily been considered to be "failure," that is, the lack of ever achieving the *terra firma* of moral certitude, is reconceived of as a kind of success and advancement. Consequently, "failure," Žižek argues, "is no longer perceived as opposed to success, since success itself can consist only in heroically assuming the full dimension of failure itself, 'repeating' failure as 'one's own'" (ibid.). In other words, the provisory nature of ethics is not a failure as opposed to some other presumed outcome that would be called "success." Instead, it is only by assuming and affirming this supposed "failure" that what is called ethics will have succeeded.

In stating this, however, we immediately run up against the so-called problem of *relativism*—"the claim that no universally valid beliefs or values exist" (Ess 1996, 204). To put it directly, if all morality is provisional and open to different decisions concerning difference made at different times for different reasons, are we not at risk of affirming an extreme form of moral relativism? We should respond to this indictment not by seeking some definitive and universally accepted response (which would obviously answer the charge of relativism by taking refuge in and validating its opposite), but by following Žižek's (2000, 3) strategy of "fully endorsing what one is accused of." So yes, relativism, but an extreme and carefully articulated version. That is, a relativism that can no longer be comprehended by that kind of understanding of the term which makes it the mere negative and counterpoint of universalism. This understanding of "relative" would, therefore, be similar to what has been developed in physics beginning with Albert Einstein, that is, a conceptualization capable of acknowledging that everything (the terms of this statement included) is in motion and that there neither is nor can be a fixed point from which to observe or evaluate anything. Or to put it in Cartesian language, any decision concerning a "fixed point" would have to be and would remain *provisional*. Understood in this way, then, relativism is not the mere polar opposite of universalism but the ground (which is, of course, no "ground" in the usual sense of the word but something like "condition for possibility") from which the terms "universal" and "relative" will have been formulated and deployed in the first place.

If what is ultimately sought and valued is a kind of morality that is locked down and secured through the metaphysical certitude provided by some transcendental figure like a god, then this outcome would be virtually indistinguishable from "plain old relativism." But once it is admitted that this conceptual anchor has been cut loose—that is, after the death or termination of all the customary moral authority figures like god in Nietzsche

(1974), the author in Barthes (1978), and the human subject in Heidegger (1977c) and Foucault (1973)—all things appear to be open to reconfiguration and reevaluation. This occurrence, as Nietzsche (1974, 279) had written concerning the "death of god," is only able to be considered a deficiency and problem from a position that always and already validated the assumption of a fixed and immovable point of view—the equivalent of a moral Ptolemaic system. But if viewed from an alternative perspective, this situation can be affirmed as an opening and dynamic opportunity. In Nietzsche's words: "And these initial consequences, the consequences for ourselves, are quite the opposite of what one might perhaps expect: They are not at all sad and gloomy but rather like a new and scarcely describable kind of light, happiness, relief, exhilaration, encouragement, dawn" (ibid., 280).

Relativism, therefore, does not necessarily need to be construed negatively and decried, as Žižek (2003, 79; 2006a, 281) has often done, as the epitome of postmodern multiculturalism run amok. It too can and should be understood otherwise. Robert Scott, for instance, understands "relativism" to be otherwise than a pejorative term: "Relativism, supposedly, means a standardless society, or at least a maze of differing standards, and thus a cacophony of disparate, and likely selfish, interests. Rather than a standardless society, which is the same as saying no society at all, relativism indicates circumstances in which standards have to be established cooperatively and renewed repeatedly" (Scott 1967, 264). Or as James Carey describes it in his seminal essay "A Cultural Approach to Communication": "All human activity is such an exercise in squaring the circle. We first produce the world by symbolic work and then take up residence in the world we have produced. Alas, there is magic in our self deceptions" (Carey 1989, 30).

In fully endorsing this form of relativism and following through on it to the end, what one gets is not necessarily what might have been expected, namely, a situation where anything goes and "everything is permitted" (Camus 1983, 67). Instead, what is obtained is a kind of ethical thinking that turns out to be much more responsive and responsible. Ethics, conceived of in this way, is about decision and not discovery (Putnam 1964, 691). *We*, individually and in collaboration with each other (and not just those others who we assume are substantially like ourselves), decide who is and who is not part of the moral community—who, in effect, will have been admitted to and included in this first-person plural pronoun. This decision, as Anne Foerst (Benford and Malartre 2007, 163) points out, is never certain; it is always and continues to be provisional. In effect, and to paraphrase Carey, we make the rules for ourselves and those we consider Other and then play by them . . . or not.

Should machines like AIs, robots, and other autonomous systems be granted admission to the community of moral subjects, becoming what would be recognized as legitimate moral agents, patients, or both? This question cannot be answered definitively and finally with a simple "yes" or "no." The question will need to be asked and responded to repeatedly in specific circumstances. But the question needs to be asked and explicitly addressed rather than being passed over in silence as if it did not matter. As Norbert Wiener predicted over a half-century ago, "Society can only be understood through a study of the messages and the communication facilities which belong to it; and . . . in the future development of these messages and communication facilities, messages between man and machines, between machines and man, and between machine and machine, are destined to play an ever increasing part" (Wiener 1954, 16). What matters, then, is how one responds, how the terms and conditions of these relationships are decided, and how responsibility comes to be articulated in the face of all these others.

Consequently, we, and we alone, are responsible for determining the scope and boundaries of moral responsibility, for instituting these decisions in everyday practices, and for evaluating their results and outcomes. We are, in effect, responsible for deciding who or what is to be included in this "we" and who or what is not. Although we have often sought to deflect these decisions and responsibilities elsewhere, typically into the heavens but also on other terrestrial authorities, in order to validate and/ or to avoid having to take responsibility for them, we are, in the final analysis, the sole responsible party. It is a *fraternal logic*, but one for which we must take full responsibility. This means of course that whoever is empowered to make these decisions must be vigilant and critical of the assignments that are made, who or what comes to be included and why, who or what remains excluded and why, and what this means for us, for others, and the subject of ethics. And, as Calarco (2008, 77) points out, there are "no guarantees that we have gotten things right." Mistakes and missteps are bound to happen. What matters, however, is that we take full responsibility for these failures rather than making excuses by way of deflection or deferral to some transcendental authority or universal values. We are, therefore, not just responsible for acting responsibly in accordance with ethics; we are responsible for ethics. In other words, the machine is not just another kind of other who calls to us and requires a suitable moral response. The machine puts "the questioning of the other" (Levinas 1969, 178) into question and asks us to reconsider without end "what respond means" (Derrida 2008, 8).

Notes

Introduction

1. This concept of "pure agency," although excluded from further consideration by Floridi and Sanders, will turn out to be operationalized by functionalist approaches to designing artificial autonomous agents (AAAs). This development will be explicitly analyzed in the consideration of machine moral agency.

1   Moral Agency

1. The fact that it is not explicitly identified as such could be taken as evidence of the extent to which the instrumental definition has become so widely accepted and taken for granted as to be virtually transparent.

2. In her later work, Johnson has increasingly recognized the complexity of agency in situations involving advanced computer systems. "When computer systems behave," she writes in the essay "Computer Systems: Moral Entities but Not Moral Agents," "there is a triad of intentionality at work, the intentionality of the computer system designer, the intentionality of the system, and the intentionality of the user" (Johnson 2006, 202). Although this statement appears to complicate the human-centric perspective of computer ethics and allow for a more distributed model of moral agency, Johnson still insists on the privileged status and position of the human subject: "Note also that while human beings can act with or without artifacts, computer systems cannot act without human designers and users. Even when their proximate behavior is independent, computer systems act with humans in the sense that they have been designed by humans to behave in certain ways and humans have set them in particular places, at particular times, to perform particular tasks for users" (ibid.).

3. Although Kant, unlike his predecessors, Descartes and Leibniz in particular, does not give serious consideration to the possibility of the *machina ratiocinatrix*, he does, in the *Anthropology* (2006), entertain the possibility of otherworldly nonhuman rational beings, that is, extraterrestrials or space aliens. See David Clark's "Kant's Aliens" (2001) for a critical investigation of this material.

4. The role of human responsibility in this matter would then be more complicated. It would not be a question of whether human designers and operators use the object in a way that is responsible; rather, as Bechtel (1985, 297) describes it, "the programmer will bear responsibility for preparing these systems to take responsibility." Or as Stahl (2006, 212) concludes, rephrasing the question of computer responsibility by referring it elsewhere, "can (or should) man assume the responsibility for holding computers (quasi-)responsible?"

5. There is something of an ongoing debate concerning this issue between John Cottingham and Tom Regan. Cottingham, one of Descartes's translators and Anglophone interpreters, argues, in direct opposition to Regan and Singer's *Animal Rights and Human Obligations* (1976), that animal rights philosophers have unfortunately employed a misconstrued version of Cartesian philosophy. "The standard view," Cottingham (1978, 551) writes, "has been reiterated in a recent collection on animal rights [Regan and Singer 1976], which casts Descartes as the villain of the piece for his alleged view that animals merely behave '*as if* they fell pain when they are, say, kicked or stabbed.' . . . But if we look at what Descartes actually says about animals it is by no means clear that he holds the monstrous view which all the commentators attribute to him." In response to this, Regan (1983, 4) partially agrees: "Cottingham, then, is correct to note that, as in his letter to More, Descartes does not deny that animals have sensations; but he is incorrect in thinking, as he evidently does, that Descartes thinks that animals are conscious."

6. This is already evident in Descartes's text, in which the terms "soul" and "mind" are used interchangeably. In fact, the Latin version of the *Meditations* distinguishes between "mind" and "body" whereas the French version of the same text uses the terms "soul" and "body" (Descartes 1988, 110).

7. Floridi and Sanders provide a more detailed account of "the method of abstraction" in their paper "The Method of Abstraction" (2003).

8. In taking mathematics as a model for revising and introducing conceptual rigor into an area of philosophy that has lacked such precision, Floridi and Sanders (2004) deploy one of the defining gestures of modern philosophy from Descartes to Kant but obviously extending as far back as Plato and into contemporary analytic thought.

9. There is a certain intellectual attraction to repositioning Immanuel Kant as an engineer. For instance, the *First Critique* is, as Kant had described it, nothing less than an attempt to reengineer philosophy in order to make it function more effectively and efficiently. "In fact," Alistair Welchman (1997, 218) argues, "the critical works represent a close collaboration of the traditional dogmatic understanding of engineering as mere application, of the machine as mere instrument and of matter as mere patient. . . . But this also ties closely into a thought of transcendental production that is dogmatically machinic, and engages Kant in a series of problems that are recognizable as engineering problems but that are also insoluble given the subordination of engineering to science."

10. Initially, beginning with the 1942 short story "Runaround" (Asimov, 2008), there were three laws. In later years, especially the 1985 novel *Robots and Empire*, Asimov modified the list by adding what he termed the zeroth law, "A robot may not harm humanity, or, by inaction, allow humanity to come to harm." This addition took the number zero as opposed to four in order to retain the hierarchical cascading structure where lower-numbered laws had precedence over those with higher-numbers.

2 Moral Patiency

1. The term "moral patient" has been used in analytic ethics' engagement with the animal question. Continental philosophers generally do not use the term but talk instead about "the other" and "otherness." This difference is not simply a nominal issue. It will turn out to be a crucial and important one.

2. Birch's use of the Latin prefix *homo-* as opposed to the Greek *anthropo-* provides for an interesting effect, when read across the two languages. *Homo* in Latin means "man," but in Greek the same word means "the same." Consequently, Birch's "*homo-centric* ethics," names both a form of ethics that is more of the same insofar as it is and remains centered on an exclusively human subject.

3. The choice of Chinese in this illustration is neither accidental nor unprecedented. Obviously any other language could have been employed. Why Chinese? This is because the Chinese language and its written characters in particular comprise, for the modern European imagination, the principal figure of otherness. Leibniz in particular was fascinated with Chinese writing and positioned it as a conceptual counterpoint to European languages and script (Derrida 1976, 79–80).

4. For more on both *To Tell the Truth* and *What's My Line*, see what is arguably the definitive resource for information regarding popular culture and related phenomena, Wikipedia (http://en.wikipedia.org/wiki/To_Tell_the_Truth).

5. A similar Levinasian influenced position is advanced by Silva Benso in *The Face of Things*. "Ethics," Benso (2000, 131) writes, "does not deal primarily with being good, bad, or evil. Rather, it deals with how much of reality one is able to maintain—not necessarily the ontological reality available to human beings, but rather the metaphysical reality, the other reality subtracting itself to conceptualization. What is good is defined then in terms of what preserves the maximum of reality from destruction, whereas what is bad is what works against reality, for its destruction and annihilation. The *metron* of ethics becomes not an abstract principle of value, but reality itself, its concreteness, the gravity of things." Although not using the same terminology, what Benso calls "destruction" looks substantially similar to what Floridi terms "entropy." These points of intersection and connection, however, are often missed and left unaddressed due to the analytic/continental divide that persists within the discipline of philosophy. If ever there was a time and a reason for

opening a sustained and productive dialog (as opposed to dismissive reaction or mere indifferent toleration) between the these two parties, it is for and in the face of this alternative thinking of ethics.

6. This is perhaps best illustrated by a 2008 special edition of *Ethics and Information Technology* 10(2–3) titled "Luciano Floridi's Philosophy of Information and Information Ethics: Critical Reflections and the State of the Art" and edited by Charles Ess.

3 Thinking Otherwise

1. This analytic moniker is not ever used by Levinas, who is arguably the most influential moral thinker in the continental tradition. The term is, however, employed by a number of Levinas's Anglophone interpreters. Simon Critchley (2002, 25), for instance, utilizes "the old epistemological chestnut of the problem of other minds" in an effort to explain the approach and importance of Levinasian thought. Likewise Adriaan Peperzak (1997, 33) makes reference to the "theoreticians of the 'other minds' problematic" in order to situate Levinas's philosophy as fundamentally different in its approach to and understanding of difference.

2. I realize that employing the term "deconstruction" in this particular context is somewhat problematic. This is because deconstruction does not necessarily sit well with Levinas's own work. Levinas, both personally and intellectually, had a rather complex relationship with the writings of Jacques Derrida, the main proponent of what is often mislabeled "deconstructivism," and an even more complicated if not contentious one with Martin Heidegger, the thinker who Derrida credits with having first introduced the concept and practice.

3. Cohen's essay "Ethics and Cybernetics: Levinasian Reflections" was initially composed for and presented at the conference Computer Ethics: A Philosophical Exploration, held at the London School of Economics and Political Science on December 14–15, 1998. It was first published in the journal *Ethics and Information Technology* in 2000 and subsequently reprinted in Peter Atterton and Matthew Calarco's *Radicalizing Levinas* (2010).

4. For a detailed consideration of the history of the concept and the development of the science, see N. Katherine Hayles's *How We Became Posthuman* (1999). This text not only provides a critical analysis of the evolution of cybernetics, including a detailed consideration of its three historical epochs or "waves" of homeostasis (1945–1960), reflexivity (1960–1980), and virtuality (1980–1999) (Hayles 1999, 7); the role of the Macy Conferences of Cybernetics, which were held from 1943 to 1954; and the major figures involved in each iteration, e.g., Norbert Wiener, Claude Shannon, Warren McCulloch, Margaret Mead, Gregory Bateson, Heinz von Foerster, Humberto Maturana, and Francisco Varela; but also establishes the foundation and protocols for what is arguably a "fourth wave," where cybernetics has been repurposed for what is now called, in the wake of Donna Haraway's (1991)

groundbreaking work, the "posthumanities." For more on this development, see Carey Wolfe's *What Is Posthumanism?* (2010).

5. The subtitle to Levinas's (1969) *Totality and Infinity* is "An Essay on Exteriority."

6. This emphasis on "excess" and the "excessive," it should be noted, is both deliberate and entirely appropriate. Some of the earliest texts of philosophy, at least as Socrates describes it in the *Protagoras*, appear in the temple at Delphi in the form of two laconic imperatives: γνῶθι σεαυτόν, "Know Thy Self," and μηδέν ἄγαν, "Nothing in Excess" (Plato 1977, 343b). Typically, these statements have been read as moral directives or maxims. The first specifies that one should seek to attain self-knowledge. That is, the "lover of wisdom" should pursue knowledge not only of things but knowledge of the mode of the knowing of things—a self-aware and self-reflective understanding that makes one's own method of knowing things an issue for oneself. The second is usually interpreted as a prohibition, stipulating that everything, presumably even this self-knowledge, should be undertaken and pursued within the right measure. Nothing should, so it is suggested, be taken to an extreme; everything should be contained within its proper limits and boundaries. There is, however, another way to read these statements that provides for an alternative account and perspective. In particular, the second statement may be read ontologically rather than as a proscription. In this way, "nothing in excess" signifies that whatever exceeds the grasp of the self's self-knowing—that is, whatever resists and falls outside the capabilities and scope of "know thy self"—will have been no-thing. In other words, whatever exceeds the grasp of self-knowing and resides on the exterior of this particular kind of knowledge will have been regarded as nothing and considered of no consequence. (And is this not precisely what Descartes had proposed as a consequence of the *cogito ergo sum*?) Consequently, it is this operation, this decisive cut, that institutes and normalizes exclusion. Philosophy, right at the very beginning, with its attention to the Delphic Oracle, decides on an exclusive undertaking, turning everything and anything that exceeds the grasp of philosophical self-knowledge into nothing.

7. Although Hegel was no computer scientist, his concept of the "bad or spurious infinite" is remarkably similar to "recursion," a fundamental aspect of computational operations that defines an infinite number of instances by using a finite expression. A similar concept is also articulated by Niklas Luhmann (1995, 9), who argued that "the distinction between 'closed' and 'open' systems is replaced by the question of how self-referential closure can create openness."

References

Note: All documented URLs valid as of February 2012.

Achebe, Chinua. 1994. *Things Fall Apart*. New York: Anchor Books.

Adam, Alison. 2008. Ethics for things. *Ethics and Information Technology* 10:149–154.

Allen, Colin, Iva Smit, and Wendell Wallach. 2006. Artificial morality: Top-down, bottom-up, and hybrid approaches. *Ethics and Information Technology* 7:149–155.

Allen, Colin, Gary Varner, and Jason Zinser. 2000. Prolegomena to any future artificial moral agent. *Journal of Experimental & Theoretical Artificial Intelligence* 12:251–261.

Allen, Colin, Wendell Wallach, and Iva Smit. 2006. Why machine ethics? *IEEE Intelligent Systems* 21 (4):12–17.

Anderson, Michael, and Susan Leigh Anderson. 2006. Machine ethics. *IEEE Intelligent Systems* 21 (4):10–11.

Anderson, Michael, and Susan Leigh Anderson. 2007a. Machine ethics: Creating an ethical intelligent agent. *AI Magazine* 28 (4):15–26.

Anderson, Michael, and Susan Leigh Anderson. 2007b. The status of machine ethics: A report from the AAAI Symposium. *Minds and Machines* 17 (1):1–10.

Anderson, Michael, Susan Leigh Anderson, and Chris Armen. 2004. Toward machine ethics. *American Association for Artificial Intelligence*. http://www.aaai.org/Papers/Workshops/2004/WS-04-02/WS04-02-008.pdf.

Anderson, Michael, Susan Leigh Anderson, and Chris Armen. 2006. An approach to computing ethics. *IEEE Intelligent Systems* 21 (4):56–63.

Anderson, Susan Leigh. 2008. Asimov's "Three Laws of Robotics" and machine metaethics. *AI & Society* 22 (4):477–493.

Apel, Karl-Otto. 2001. *The Response of Discourse Ethics*. Leuven, Belgium: Peeters.

Arrabales, Raul, Agapito Ledezma, and Araceli Sanchis. 2009. Establishing a roadmap and metric for conscious machines development. Paper presented at the 8th IEEE International Conference on Cognitive Informatics, Hong Kong, China, June 15–17. http://www.conscious-robots.com/raul/papers/Arrabales_ICCI09_preprint.pdf.

Asaro, Peter M. 2007. Robots and responsibility from a legal perspective. In *Proceedings of the IEEE Conference on Robotics and Automation, Workshop on Roboethics*. Rome, Italy, April 14. http://www.cybersophe.org/writing/ASARO%20Legal%20Perspective .pdf.

Asimov, Isaac. 1976. *The Bicentennial Man and Other Stories*. New York: Doubleday.

Asimov, Isaac. 1983. *Asimov on Science Fiction*. New York: HarperCollins.

Asimov, Isaac. 1985. *Robots and Empire*. New York: Doubleday.

Asimov, Isaac. 2008. *I, Robot*. New York: Bantam Books.

Atterton, Peter, and Matthew Calarco, eds. 2010. *Radicalizing Levinas*. Albany, NY: SUNY Press.

Augustine, Saint. 1963. *The Confessions of St. Augustine*. Trans. Rex Warner. New York: New American Library.

Balluch, Martin, and Eberhart Theuer. 2007. Trial on personhood for chimp "Hiasl." *ALTEX* 24 (4):335–342. http://www.vgt.at/publikationen/texte/artikel/20080118Hiasl .htm.

Balthasar, Hans Urs von. 1986. On the concept of person. *Communio: International Catholic Review* 13 (spring):18–26.

Barthes, Roland. 1978. *Image, Music, Text*. Trans. Stephen Heath. New York: Hill & Wang.

Bates, J. 1994. The role of emotion in believable agents. *Communications of the ACM* 37:122–125.

Bateson, M. 2004. Mechanisms of decision-making and the interpretation of choice tests. *Animal Welfare* 13 (supplement):S115–S120.

Bateson, P. 2004. Do animals suffer like us? *Veterinary Journal* 168:110–111.

Battlestar Galactica. 2003–2009. NBC Universal Pictures.

Bayley, Barrington J. 1974. *The Soul of the Robot*. Gillette, NJ: Wayside Press.

Beauchamp, T. L., and J. F. Childress. 1979. *Principles of Biomedical Ethics*. Oxford: Oxford University Press.

Bechtel, W. 1985. Attributing responsibility to computer systems. *Metaphilosophy* 16 (4):296–305.

Bell, Charles. 1806. *The Anatomy and Philosophy of Expression: As Connected with the Fine Arts*. London: R. Clay, Son & Taylor.

Benford, Gregory, and Elisabeth Malartre. 2007. *Beyond Human: Living with Robots and Cyborgs*. New York: Tom Doherty.

Benso, Silvia. 2000. *The Face of Things: A Different Side of Ethics*. Albany, NY: SUNY Press.

Bentham, Jeremy. 2005. *An Introduction to the Principles of Morals and Legislation*. Ed. J. H. Burns and H. L. Hart. Oxford: Oxford University Press.

Birch, Thomas H. 1993. Moral considerability and universal consideration. *Environmental Ethics* 15:313–332.

Birch, Thomas H. 1995. The incarnation of wilderness: Wilderness areas as prisons. In *Postmodern Environmental Ethics*, ed. Max Oelschlaeger, 137–162. Albany, NY: SUNY Press.

Birsch, Douglas. 2004. Moral responsibility for harm caused by computer system failures. *Ethics and Information Technology* 6:233–245.

Blackmore, S. 2003. *Consciousness: An Introduction*. London: Hodder & Stoughton.

Block, Ned Joel, Owen J. Flanagan, and Güven Güzeldere. 1997. *The Nature of Consciousness: Philosophical Debates*. Cambridge, MA: MIT Press.

Blumberg, B., P. Todd, and M. Maes. 1996. No bad dogs: Ethological lessons for learning. In *Proceedings of the 4th International Conference on Simulation of Adaptive Behavior* (SAB96), 295–304. Cambridge, MA: MIT Press.

Boethius. 1860. *Liber de persona et duabus naturis contra Eutychen et Nestorium, ad Joannem Diaconum Ecclesiae Romanae*: c. iii (*Patrologia Latina* 64). Paris.

Bostrom, Nick. 2003. Ethical issues in advanced artificial intelligence. In *Cognitive, Emotive and Ethical Aspects of Decision Making in Humans and Artificial Intelligence*, vol. 2, ed. Iva Smit, Wendell Wallach, and George E. Lasker, 12–17. International Institute for Advanced Studies in Systems Research and Cybernetics. http://www.nickbostrom.com/ethics/ai.pdf.

Breazeal, Cynthia, and Rodney Brooks. 2004. Robot Emotion: A Functional Perspective. In *Who Needs Emotions: The Brain Meets the Robot*, ed. J. M. Fellous and M. Arbib, 271–310. Oxford: Oxford University Press.

Brey, Philip. 2008. Do we have moral duties towards information objects? *Ethics and Information Technology* 10:109–114.

Bringsjord, Selmer. 2006. Toward a general logicist methodology for engineering ethically correct robots. *IEEE Intelligent Systems* 21 (4):38–44.

Bringsjord, Selmer. 2008. Ethical robots: The future can heed us. *AI & Society* 22:539–550.

Brooks, Rodney A. 1999. *Cambrian Intelligence: The Early History of the New AI*. Cambridge, MA: MIT Press.

Brooks, Rodney A. 2002. *Flesh and Machines: How Robots Will Change Us*. New York: Pantheon Books.

Butler, Judith. 1990. *Gender Trouble: Feminism and the Subversion of Idenity*. New York: Routledge.

Bryson, Joanna. 2010. Robots should be slaves. In *Close Engagements with Artificial Companions: Key Social, Psychological, Ethical and Design Issues*, ed. Yorick Wilks, 63–74. Amsterdam: John Benjamins.

Calarco, Matthew. 2008. *Zoographies: The Question of the Animal from Heidegger to Derrida*. New York: Columbia University Press.

Calverley, David J. 2006. Android science and animal rights: Does an analogy exist? *Connection Science* 18 (4):403–417.

Calverley, David J. 2008. Imaging a non-biological machine as a legal person. *AI & Society* 22:523–537.

Camus, Albert. 1983. *The Myth of Sisyphus, and Other Essays*. Trans. Justin O'Brien. New York: Alfred A. Knopf.

Čapek, Karel. 2008. *R.U.R. and The Robber: Two Plays by Karl Čapek*. Ed. and trans. Voyen Koreis. Brisbane: Booksplendour Publishing.

Capurro, Rafael. 2008. On Floridi's metaphysical foundation of information ecology. *Ethics and Information Technology* 10:167–173.

Carey, James. 1989. *Communication as Culture: Essays on Media and Society*. New York: Routledge.

Carrithers, Michael, Steven Collins, and Steven Lukes, eds. 1985. *The Category of the Person: Anthropology, Philosophy, History*. Cambridge: Cambridge University Press.

Chalmers, David J. 1996. *The Conscious Mind: In Search of a Fundamental Theory*. New York: Oxford University Press.

Channell, David F. 1991. *The Vital Machine: A Study of Technology and Organic Life*. Oxford: Oxford University Press.

Cherry, Christopher. 1991. Machines as persons? In *Human Beings*, ed. David Cockburn, 11–24. Cambridge: Cambridge University Press.

Chopra, Samir, and Laurence White. 2004. Moral agents—Personhood in law and philosophy. In *Proceedings from the European Conference on Artificial Intelligence (ECAI)*,

August 2004 Valencia, Spain, ed. Ramon López de Mántaras and Lorena Saitta, 635–639. Amsterdam: IOS Press.

Christensen, Bill. 2006. Asimov's first law: Japan sets rules for robots. *LiveScience* (May 26). http://www.livescience.com/10478-asimov-law-japan-sets-rules-robots.html.

Churchland, Paul M. 1999. *Matter and Consciousness,* rev. ed. Cambridge, MA: MIT Press.

Clark, David. 2001. Kant's aliens: The Anthropology and its others. *CR* 1 (2):201–289.

Clark, David. 2004. On being "the last Kantian in Nazi Germany": Dwelling with animals after Levinas. In *Postmodernism and the Ethical Subject,* ed. Barbara Gabriel and Suzan Ilcan, 41–74. Montreal: McGill-Queen's University Press.

Coeckelbergh, Mark. 2010. Moral appearances: Emotions, robots, and human morality. *Ethics and Information Technology* 12 (3):235–241.

Cohen, Richard A. 2000. Ethics and cybernetics: Levinasian reflections. *Ethics and Information Technology* 2:27–35.

Cohen, Richard A. 2001. *Ethics, Exegesis, and Philosophy: Interpretation After Levinas.* Cambridge: Cambridge University Press.

Cohen, Richard A. 2003. Introduction. In *Humanism of the Other,* by Emmanuel Levinas, vii–xliv. Urbana, IL: University of Illinois Press.

Cohen, Richard A. 2010. Ethics and cybernetics: Levinasian reflections. In *Radicalizing Levinas,* ed. Peter Atterton and Matthew Calarco, 153–170. Albany, NY: SUNY Press.

Cole, Phillip. 1997. Problem with "persons." *Res Publica* 3 (2):165–183.

Coleman, Kari Gwen. 2001. Android arete: Toward a virtue ethic for computational agents. *Ethics and Information Technology* 3:247–265.

Coman, Julian. 2004. Derrida, philosophy's father of "deconstruction," dies at 74. *Telegraph,* October 10. http://www.telegraph.co.uk/news/worldnews/europe/france/1473821/Derrida-philosophys-father-of-deconstruction-dies-at-74.html.

Computer Ethics Institute. 1992. Ten commandments of computer ethics. http://computerethicsinstitute.org.

Copland, Jack. 2000. What is artificial intelligence. AlanTuring.net. http://www.alanturing.net/turing_archive/pages/Reference%20Articles/what_is_AI/What%20is%20AI09.html.

Cottingham, John. 1978. A brute to the brutes? Descartes's treatment of animals. *Philosophy* 53 (206):551–559.

Critchley, Simon. 2002. Introduction. In *The Cambridge Companion to Levinas,* ed. Simon Critchley and Robert Bernasconi, 1–32. Cambridge: Cambridge University Press.

Danielson, Peter. 1992. *Artificial Morality: Virtuous Robots for Virtual Games*. New York: Routledge.

Darwin, Charles. 1998. *The Expression of the Emotions in Man and Animals*. Oxford: Oxford University Press. Originally published 1872.

Dawkins, Marian Stamp. 2001. Who needs consciousness? *Animal Welfare* 10:319–329.

Dawkins, Marian Stamp. 2008. The science of animal suffering. *Ethology* 114 (10):937–945.

DeGrazia, David. 2006. On the question of personhood beyond Homo sapiens. In *Defense of Animals: The Second Wave*, ed. Peter Singer, 40–53. Oxford: Blackwell.

Deleuze, Gilles. 1994. *Difference and Repetition*. Trans. Paul Patton. New York: Columbia University Press.

Dennett, Daniel C. 1989. *The Intentional Stance*. Cambridge, MA: MIT Press.

Dennett, Daniel C. 1994. The practical requirements for making a conscious robot. *Philosophical Transactions of the Royal Society* A349:133–146.

Dennett, Daniel C. 1996. *Kinds of Minds: Toward an Understanding of Consciousness*. New York: Basic Books.

Dennett, Daniel C. 1997. When HAL kills, who's to blame? Computer ethics. In *Hal's Legacy: 2001's Computer as Dream and Reality*, ed. David G. Stork, 351–366. Cambridge, MA: MIT Press.

Dennett, Daniel C. 1998. *Brainstorms: Philosophical Essays on Mind and Psychology*. Cambridge, MA: MIT Press.

Derrida, Jacques. 1973. *Speech and Phenomena*. Trans. David B. Allison. Evanston, IL: Northwestern University Press.

Derrida, Jacques. 1976. *Of Grammatology*. Trans. Gayatri Chakravorty Spivak. Baltimore, MD: The Johns Hopkins University Press.

Derrida, Jacques. 1978. *Writing and Difference*. Trans. Alan Bass. Chicago: University of Chicago Press.

Derrida, Jacques. 1981. *Positions*. Trans. Alan Bass. Chicago: University of Chicago Press.

Derrida, Jacques. 1982. *Margins of Philosophy*. Trans. Alan Bass. Chicago: University of Chicago Press.

Derrida, Jacques. 1988. *Limited Inc*. Trans. Samuel Weber. Evanston, IL: Northwestern University Press.

Derrida, Jacques. 2005. *Paper Machine*. Trans. Rachel Bowlby. Stanford, CA: Stanford University Press.

Derrida, Jacques. 2008. *The Animal That Therefore I Am*. Ed. Marie-Louise Mallet. Trans. David Wills. New York: Fordham University Press.

Descartes, Rene. 1988. *Selected Philosophical Writings*. Trans. John Cottingham, Robert Stoothoff, and Dugald Murdoch. Cambridge: Cambridge University Press.

Dick, Philip K. 1982. *Do Androids Dream of Electric Sheep?* New York: Ballantine Books.

DJ Danger Mouse. 2004. *The Grey Album*. Self-released.

Dodig-Crnkovic, Gordana, and Daniel Persson. 2008. Towards trustworthy intelligent robots—A pragmatic approach to moral responsibility. Paper presented to the North American Computing and Philosophy Conference, NA-CAP@IU 2008. Indiana University, Bloomington, July 10–12. http://www.mrtc.mdh.se/~gdc/work/NACAP-Roboethics-Rev1.pdf.

Dolby, R. G. A. 1989. The possibility of computers becoming persons. *Social Epistemology* 3 (4):321–336.

Donath, Judith. 2001. Being real: Questions of tele-identity. In *The Robot in the Garden: Telerobotics and Telepistemology in the Age of the Internet*, ed. Ken Goldberg, 296–311. Cambridge, MA: MIT Press.

Dracopoulou, Souzy. 2003. The ethics of creating conscious robots—life, personhood, and bioengineering. *Journal of Health, Social and Environmental Issues* 4 (2):47–50.

Dumont, Étienne. 1914. *Bentham's Theory of Legislation*. Trans. Charles Milner Atkinson. Oxford: Oxford University Press.

Ellul, Jacques. 1964. *The Technological Society*. Trans. John Wilkinson. New York: Vintage Books.

Ess, Charles. 1996. The political computer: Democracy, CMC, and Habermas. In *Philosophical Perspectives on Computer-Mediated Communication*, ed. Charles Ess, 196–230. Albany, NY: SUNY Press.

Ess, Charles. 2008. Luciano Floridi's philosophy of information and information ethics: Critical reflections and the state of the art. *Ethics and Information Technology* 10:89–96.

Feenberg, Andrew. 1991. *Critical Theory of Technology*. Oxford: Oxford University Press.

Floridi, Luciano. 1999. Information ethics: On the philosophical foundation of computer ethics. *Ethics and Information Technology* 1 (1):37–56.

Floridi, Luciano. 2002. On the intrinsic value of information objects and the infosphere. *Ethics and Information Technology* 4:287–304.

Floridi, Luciano. 2003. Two approaches to the philosophy of information. *Minds and Machines* 13:459–469.

Floridi, Luciano. 2004. Open problems in the philosophy of information. *Metaphilosophy* 35 (4):554–582.

Floridi, Luciano. 2008. Information ethics: Its nature and scope. In *Information Technology and Moral Philosophy*, ed. Jeroen van den Hoven and John Weckert, 40–65. Cambridge: Cambridge University Press.

Floridi, Luciano. 2010. Information ethics. In *Cambridge Handbook of Information and Computer Ethics*, ed. Luciano Floridi, 77–100. Cambridge: Cambridge University Press.

Floridi, Luciano, and J. W. Sanders. 2001. Artificial evil and the foundation of computer ethics. *Ethics and Information Technology* 3 (1):56–66.

Floridi, Luciano, and J. W. Sanders. 2003. The method of abstraction. In *Yearbook of the Artificial: Nature, Culture, and Technology: Models in Contemporary Sciences*, 117–220. Bern: Peter Lang. http://citeseerx.ist.psu.edu/viewdoc/download?doi=10.1.1.66 .3827&rep=rep1&type=pdf.

Floridi, Luciano, and J. W. Sanders. 2004. On the morality of artificial agents. *Minds and Machines* 14:349–379.

Foucault, Michel. 1973. *The Order of Things: An Archaeology of the Human Sciences*. Trans. Alan Sheridan. New York: Vintage Books.

Franklin, Stan. 2003. IDA: A conscious artifact? In *Machine Consciousness*, ed. Owen Holland, 47–66. Charlottesville, VA: Imprint Academic.

French, Peter. 1979. The corporation as a moral person. *American Philosophical Quarterly* 16 (3):207–215.

Freitas, Robert A. 1985. The legal rights of robots. *Student Lawyer* 13:54–56.

Friedenberg, Jay. 2008. *Artificial Psychology: The Quest for What It Means to Be Human*. New York: Taylor & Francis.

Georges, Thomas M. 2003. *Digital Soul: Intelligent Machines and Human Values*. Boulder, CO: Westview Press.

Gibson, William. 2005. God's little toys: Confessions of a cut and paste artist. *Wired* 13 (7):118–119.

Gizmodo. 2010. Shimon robot takes over jazz as doomsday gets a bit more musical. http://gizmodo.com/5228375/shimon-robot-takes-over-jazz-as-doomsday -gets-a-bit-more-musical.

Godlovitch, Stanley, Roslind Godlovitch, and John Harris, eds. 1972. *Animals, Men and Morals: An Enquiry into the Maltreatment of Non-humans*. New York: Taplinger Publishing.

Goertzel, Ben. 2002. Thoughts on AI morality. *Dynamical Psychology: An International, Interdisciplinary Journal of Complex Mental Processes* (May). http://www.goertzel .org/dynapsyc/2002/AIMorality.htm.

Goodpaster, Kenneth E. 1978. On being morally considerable. *Journal of Philosophy* 75:303–325.

Grau, Christopher. 2006. There is no "I" in "robot": Robots and utilitarianism. *IEEE Intelligent Systems* 21 (4):52–55.

Grodzinsky, Frances S., Keith W. Miller, and Marty J. Wolf. 2008. The ethics of designing artificial agents. *Ethics and Information Technology* 10:115–121.

Guarini, Marcello. 2006. Particularism and the classification and reclassification of moral cases. *IEEE Intelligent Systems* 21 (4):22–28.

Gunkel, David J. 2007. *Thinking Otherwise: Philosophy, Communication, Technology.* West Lafayette, IN: Purdue University Press.

Güzeldere, Güven. 1997. The many faces of consciousness: A field guide. In *The Nature of Consciousness: Philosophical Debates*, ed. Ned Block, Owen Flanagan, and Güven Güzeldere, 1–68. Cambridge, MA: MIT Press.

Haaparanta, Leila. 2009. *The Development of Modern Logic.* Oxford: Oxford University Press.

Habermas, Jürgen. 1998. *The Inclusion of the Other: Studies in Political Theory.* Trans. and ed. Ciaran P. Cronin and Pablo De Greiff. Cambridge, MA: MIT Press.

Habermas, Jürgen. 1999. *Moral Consciousness and Communicative Action.* Trans. Christian Lenhardt and Shierry Weber Nicholsen. Cambridge, MA: MIT Press.

Haikonen, Pentti O. 2007. *Robot Brains: Circuits and Systems for Conscious Machines.* Chichester: Wiley.

Hajdin, Mane. 1994. *The Boundaries of Moral Discourse.* Chicago: Loyola University Press.

Haley, Andrew G. 1963. *Space Law and Government.* New York: Appleton-Century-Crofts.

Hall, J. Storrs. 2001. Ethics for machines. KurzweilAI.net, July 5. http://www.kurzweilai.net/ethics-for-machines.

Hall, J. Storrs. 2007. *Beyond AI: Creating the Consciousness of the Machine.* Amherst, NY: Prometheus Books.

Hallevy, Gabriel. 2010. The criminal liability of artificial intelligent entities. *Social Science Research Network* (SSRN). http://ssrn.com/abstract=1564096.

Hanson, F. Allan. 2009. Beyond the skin bag: On the moral responsibility of extended agencies. *Ethics and Information Technology* 11:91–99.

Haraway, Donna J. 1991. *Simians, Cyborgs, and Women: The Reinvention of Nature.* New York: Routledge.

Haraway, Donna J. 2008. *When Species Meet*. Minneapolis, MN: University of Minnesota Press.

Harman, Graham. 2002. *Tool-Being: Heidegger and the Metaphysics of Objects*. Chicago: Open Court.

Harrison, Peter. 1991. Do animals feel pain? *Philosophy* 66 (255):25–40.

Harrison, Peter. 1992. Descartes on animals. *Philosophical Quarterly* 42 (167):219–227.

Haugeland, John. 1981. *Mind Design*. Montgomery, VT: Bradford Books.

Hayles, N. Katherine. 1999. *How We Became Posthuman: Virtual Bodies in Cybernetics, Literature, and Informatics*. Chicago: University of Chicago Press.

Hegel, G. W. F. 1969. *Science of Logic*. Trans. A. V. Miller. Atlantic Highlands, NJ: Humanities Press International.

Hegel, G. W. F. 1977. *Phenomenology of Spirit*. Trans. A. V. Miller. Oxford: Oxford University Press. Originally published 1801.

Hegel, G. W. F. 1986. *Jenaer Schriften 1801–1807*. Frankfurt: SuhrkampTaschenbuch Verlag.

Hegel, G. W. F. 1987. *Hegel's Logic: Being Part One of the Encyclopaedia of the Philosophical Sciences (1830)*. Trans. William Wallace. Oxford: Oxford University Press.

Hegel, G. W. F. 1988. *Hegel's Philosophy of Mind: Being Part Three of the Encyclopaedia of the Philosophical Sciences (1830)*. Trans. William Wallace. Oxford: Oxford University Press.

Heidegger, Martin. 1962. *Being and Time*. Trans. John Macquarrie and Edward Robinson. New York: Harper & Row.

Heidegger, Martin. 1967. *What Is a Thing?* Trans. W. B. Barton, Jr., and Vera Deutsch. Chicago: Henry Regnery.

Heidegger, Martin. 1971a. The origin of the work of art. In *Poetry, Language, Thought*, 15–88. Trans. Albert Hofstadter. New York: Harper & Row.

Heidegger, Martin. 1971b. The thing. In *Poetry, Language, Thought*, 163–186. Trans. Albert Hofstadter. New York: Harper & Row.

Heidegger, Martin. 1977a. The question concerning technology. In *The Question Concerning Technology and Other Essays*, 3–35. Trans. William Lovitt. New York: Harper & Row.

Heidegger, Martin. 1977b. The end of philosophy and the task of thinking. In *Martin Heidegger Basic Writings*, ed. David F. Krell, 373–392. Trans. Joan Stambaugh. New York: Harper & Row.

Heidegger, Martin. 1977c. Letter on humanism. In *Martin Heidegger Basic Writings*, ed. David F. Krell, 213–266. New York: Harper & Row.

Heidegger, Martin. 1996. *The Principle of Reason.* Trans. Reginald Lilly. Indianapolis, IN: Indiana University Press.

Heidegger, Martin. 2010. Only a god can save us: Der Spiegel Interview (1966). In *Heidegger: The Man and the Thinker,* 45–68. Ed. and trans. Thomas Sheehan. New Brunswick, NJ: Transaction Publishers.

Heim, Michael. 1998. *Virtual Realism.* Oxford: Oxford University Press.

Herzog, Werner. 1974. *Every Man for Himself and God Against All.* München: ZDF.

Himma, Kenneth Einar. 2004. There's something about Mary: The moral value of things qua information objects. *Ethics and Information Technology* 6 (3):145–195.

Himma, Kenneth Einar. 2009. Artificial agency, consciousness, and the criteria for moral agency: What properties must an artificial agent have to be a moral agent? *Ethics and Information Technology* 11 (1):19–29.

Hoffmann, Frank. 2001. The role of fuzzy logic control in evolutionary robotics. In *Fuzzy Logic Techniques for Autonomous Vehicle Navigation,* ed. Dimiter Driankov and Alessandro Saffiotti, 119–150. Heidelberg: Physica-Verlag.

Holland, Owen. 2003. Editorial introduction. In *Machine Consciousness,* ed. Owen Holland, 1–6. Charlottesville, VA: Imprint Academic.

Ihde, Don. 2000. Technoscience and the "other" continental philosophy. *Continental Philosophy Review* 33:59–74.

Ikäheimo, Heikki, and Arto Laitinen. 2007. Dimensions of personhood: Editors' introduction. *Journal of Consciousness Studies* 14 (5–6):6–16.

Introna, Lucas D. 2001. Virtuality and morality: On (not) being disturbed by the other. *Philosophy in the Contemporary World* 8 (spring):31–39.

Introna, Lucas D. 2003. The "measure of a man" and the ethics of machines. *Lancaster University Management School Working Papers.* http://eprints.lancs.ac.uk/48690/.

Introna, Lucas D. 2009. Ethics and the speaking of things. *Theory, Culture & Society* 26 (4):398–419.

Introna, Lucas D. 2010. The "measure of a man" and the ethos of hospitality: Towards an ethical dwelling with technology. *AI & Society* 25 (1):93–102.

Johnson, Barbara. 1981. Translator's introduction. In Jacques Derrida, *Dissemination,* vii–xxxiii. Chicago: University of Chicago Press.

Johnson, Deborah G. 1985. *Computer Ethics.* Upper Saddle River, NJ: Prentice Hall.

Johnson, Deborah G. 2006. Computer systems: Moral entities but not moral agents. *Ethics and Information Technology* 8:195–204.

Johnson, Deborah G., and Keith W. Miller. 2008. Un-making artificial moral agents. *Ethics and Information Technology* 10:123–133.

Kadlac, Adam. 2009. Humanizing personhood. *Ethical Theory and Moral Practice* 13 (4):421–437.

Kant, Immanuel. 1965. *Critique of Pure Reason.* Trans. Norman Kemp Smith. New York: St. Martin's Press.

Kant, Immanuel. 1983. *Grounding for the Metaphysics of Morals.* Trans. James W. Ellington. Indianapolis, IN: Hackett.

Kant, Immanuel. 1985. *Critique of Practical Reason.* Trans. Lewis White Beck. New York: Macmillan.

Kant, Immanuel. 2006. *Anthropology from a Pragmatic Point of View.* Trans. Robert B. Louden. Cambridge: Cambridge University Press.

Kerwin, Peter. 2009. The rise of machine-written journalism. *Wired.co.uk*, December 9. http://www.wired.co.uk/news/archive/2009-12/16/the-rise-of-machine -written-journalism.aspx.

Kiekegaaard, Søren. 1987. *Either/Or, Part 2.* Trans. Howard V. Hong and Edna H. Hong. Princeton, NJ: Princeton University Press.

Kokoro, L. T. D. 2009. http://www.kokoro-dreams.co.jp/.

Koch, C. 2004. *The Quest for Consciousness. A Neurobiological Approach.* Englewood, CO: Roberts.

Kubrick, Stanley (dir.). 1968. *2001: A Space Odyssey.* Hollywood, CA: Metro-Goldwyn-Mayer (MGM).

Kuhn, Thomas S. 1996. *The Structure of Scientific Revolutions.* Chicago: University of Chicago Press.

Kurzweil, Ray. 2005. *The Singularity Is Near: When Humans Transcend Biology.* New York: Viking.

Lang, Fritz. 1927. *Metropolis.* Berlin: UFA.

Lavater, Johann Caspar. 1826. *Physiognomy, or the Corresponding Analogy between the Conformation of the Features and the Ruling Passions of the Mind.* London: Cowie, Low & Company in the Poultry.

Leiber, Justin. 1985. *Can Animals and Machines Be Persons? A Dialogue.* Indianapolis, IN: Hackett.

Leibniz, Gottfried Wilhelm. 1989. *Philosophical Essays.* Trans. Roger Ariew and Daniel Garber. Indianapolis, IN: Hackett.

Leopold, Aldo. 1966. *A Sand County Almanac.* Oxford: Oxford University Press.

Levinas, Emmanuel. 1969. *Totality and Infinity: An Essay on Exteriority.* Trans. Alphonso Lingis. Pittsburgh, PA: Duquesne University.

Levinas, Emmanuel. 1987. *Collected Philosophical Papers*. Trans. Alphonso Lingis. Dordrecht: Martinus Nijhoff.

Levinas, Emmanuel. 1990. *Difficult Freedom: Essays on Judaism*. Trans. Seán Hand. Baltimore, MD: Johns Hopkins University Press.

Levinas, Emmanuel. 1996. *Emmanuel Levinas: Basic Philosophical Writings*. Ed. Adriaan T. Peperzak, Simon Critchley, and Robert Bernasconi. Bloomington, IN: Indiana University Press.

Levinas, Emmanuel. 2003. *Humanism of the Other*. Trans. Nidra Poller. Urbana, IL: University of Illinois Press.

Levy, David. 2009. The ethical treatment of artificially conscious robots. *International Journal of Social Robotics* 1 (3):209–216.

Lippit, Akira Mizuta. 2000. *Electric Animal: Toward a Rhetoric of Wildlife*. Minneapolis: University of Minnesota Press.

Llewelyn, John. 1995. *Emmanuel Levinas: The Genealogy of Ethics*. London: Routledge.

Llewelyn, John. 2010. Pursuing Levinas and Ferry toward a newer and more democratic ecological order. In *Radicalizing Levinas*, ed. Peter Atterton and Matthew Calarco, 95–112. Albany, NY: SUNY Press.

Locke, John. 1996. *An Essay Concerning Human Understanding*. Indianapolis, IN: Hackett.

Lovelace, Ada Augusta. 1842. Translation of, and notes to, Luigi F. Menabrea's Sketch of the analytical engine invented by Charles Babbage. *Scientific Memoirs* 3: 691–731.

Lovgren, Stefan. 2007. Robot code of ethics to prevent android abuse, protect humans. *National Geographic News*, March 16. http://news.nationalgeographic.com/news/pf/45986440.html.

Lucas, Richard. 2009. *Machina Ethica: A Framework for Computers as Kant Moral Persons*. Saarbrücken: VDM Verlag.

Luhmann, Niklas. 1995. *Social Systems*. Trans. John Bednarz and Dirk Baecker. Stanford, CA: Stanford University Press.

Lyotard, Jean-François. 1984. *The Postmodern Condition: A Report on Knowledge*. Trans. Geoff Bennington and Brian Massumi. Minneapolis, MN: University of Minnesota Press.

Maner, Walter. 1980. *Starter Kit in Computer Ethics*. Hyde Park, NY: Helvetia Press and the National Information and Resource Center for Teaching Philosophy. Originally self-published in 1978.

Marx, Karl. 1977. *Capital: A Critique of Political Economy*. Trans. Ben Fowkes. New York: Vintage Books.

Mather, Jennifer. 2001. Animal suffering: An invertebrate perspective. *Journal of Applied Animal Welfare Science* 4 (2):151–156.

Matthias, Andrew. 2004. The responsibility gap: Ascribing responsibility for the actions of learning automata. *Ethics and Information Technology* 6:175–183.

Mauss, Marcel. 1985. A category of the human mind: The notion of person; the notion of self. Trans. W. D. Halls. In *The Category of the Person*, ed. Michael Carrithers, Steven Collins, and Steven Lukes, 1–25. Cambridge: Cambridge University Press.

McCarthy, John. 1979. Ascribing mental qualities to machines. http://www-formal.stanford.edu/jmc/ascribing/ascribing.html.

McCauley, Lee. 2007. AI armageddon and the Three Laws of Robotics. *Ethics and Information Technology* 9 (2):153–164.

McFarland, David. 2008. *Guilty Robots, Happy Dogs: Question of Alien Minds*. Oxford: Oxford University Press.

McLaren, Bruce M. 2006. Computational models of ethical reasoning: Challenges, initial steps, and future directions. *IEEE Intelligent Systems* 21 (4):29–37.

McLuhan, Marshall. 1995. *Understanding Media: The Extensions of Man*. Cambridge, MA: MIT Press.

McPherson, Thomas. 1984. The moral patient. *Philosophy* 59 (228):171–183.

Miller, Harlan B. 1994. Science, ethics, and moral status. *Between the Species: A Journal of Ethics* 10 (1):10–18.

Miller, Harlan B., and William H. Williams. 1983. *Ethics and Animals*. Clifton, NJ: Humana Press.

Minsky, Marvin. 2006. Alienable rights. In *Thinking about Android Epistemology*, ed. Kenneth M. Ford, Clark Glymour, and Patrick J. Hayes, 137–146. Menlo Park, CA: IAAA Press.

Moor, James H. 2006. The nature, importance, and difficulty of machine ethics. *IEEE Intelligent Systems* 21 (4):18–21.

Moore, George E. 2005. *Principia Ethica*. New York: Barnes & Noble Books. Originally published 1903.

Moravec, Hans. 1988. *Mind Children: The Future of Robot and Human Intelligence*. Cambridge, MA: Harvard University Press.

Mowbray, Miranda. 2002. Ethics for bots. Paper presented at the 14th International Conference on System Research, Informatics, and Cybernetics. Baden-Baden, Germany. July 29–August 3. http://www.hpl.hp.com/techreports/2002/HPL-2002-48R1.pdf.

Mowshowitz, Abbe. 2008. Technology as excuse for questionable ethics. *AI & Society* 22:271–282.

Murdock, Iris. 2002. *The Sovereignty of Good*. New York: Routledge.

Nadeau, Joseph Emile. 2006. Only androids can be ethical. In *Thinking about Android Epistemology*, ed. Kenneth M. Ford, Clark Glymour, and Patrick J. Hayes, 241–248. Menlo Park, CA: IAAA Press.

Naess, Arne. 1995. *Ecology, Community, and Lifestyle*. Cambridge: Cambridge University Press.

Nealon, Jeffrey. 1998. *Alterity Politics: Ethics and Performative Subjectivity*. Durham, NC: Duke University Press.

Nietzsche, Friedrich. 1966. *Beyond Good and Evil*. Trans. Walter Kaufmann. New York: Vintage Books.

Nietzsche, Friedrich. 1974. *The Gay Science*. Trans. Walter Kaufmann. New York: Vintage Books.

Nietzsche, Friedrich. 1986. *Human, All Too Human*. Trans. R. J. Hollingdale. Cambridge: Cambridge University Press.

Nissenbaum, Helen. 1996. Accountability in a computerized society. *Science and Engineering Ethics* 2:25–42.

Novak, David. 1998. *Natural Law in Judaism*. Cambridge: Cambridge University Press.

O'Regan, Kevin J. 2007. How to build consciousness into a robot: The sensorimotor approach. In *50 Years of Artificial Intelligence: Essays Dedicated to the 50th Anniversary of Artificial Intelligence*, ed. Max Lungarella, Fumiya Iida, Josh Bongard, and Rolf Pfeifer, 332–346. Berlin: Springer-Verlag.

Orwell, George. 1993. *Animal Farm*. New York: Random House/Everyman's Library.

Partridge, Derek, and Yorick Wilks, eds. 1990. *The Foundations of Artificial Intelligence: A Sourcebook*. Cambridge: Cambridge University Press.

Paton, H. J. 1971. *The Categorical Imperative: A Study in Kant's Moral Philosophy*. Philadelphia, PA: University of Pennsylvania Press.

Peperzak, Adriaan T. 1997. *Beyond the Philosophy of Emmanuel Levinas*. Evanston, IL: Northerwestern University Press.

Petersen, Stephen. 2006. The ethics of robot servitude. *Journal of Experimental & Theoretical Artificial Intelligence* 19 (1):43–54.

Plato. 1977. *Protagoras*. Trans. W. R. M. Lamb. Cambridge, MA: Harvard University Press.

Plato. 1982. *Phaedrus*. Trans. Harold North Fowler. Cambridge, MA: Harvard University Press.

Plato. 1990. *Apology*. Trans. Harold North Fowler. Cambridge, MA: Harvard University Press.

Plourde, Simmone. 2001. A key term in ethics: The person and his dignity. In *Personhood and Health Care*, ed. David C. Thomasma, David N. Weisstub, and Christian Hervé, 137–148. Dordrecht, Netherlands: Kluwer Academic Publishers.

Postman, Neil. 1993. *Technopoly: The Surrender of Culture to Technology*. New York: Vintage Books.

Powers, Thomas M. 2006. Prospects for a Kantian machine. *IEEE Intelligent Systems* 21 (4):46–51.

Putnam, Hilary. 1964. Robots: Machines or artificially created life? *Journal of Philosophy* 61 (21):668–691.

Ratliff, Evan. 2004. The crusade against evolution. *Wired* 12 (10):156–161.

Regan, Tom. 1983. *The Case for Animal Rights*. Berkeley, CA: University of California Press.

Regan, Tom. 1999. Foreword to *Animal Others: On Ethics, Ontology, and Animal Life*, ed. Peter Steeves, xi–xiii. Albany, NY: SUNY Press.

Regan, Tom, and Peter Singer, eds. 1976. *Animal Rights and Human Obligations*. New York: Prentice Hall.

Rifelj, Carol de Dobay. 1992. *Reading the Other: Novels and the Problems of Other Minds*. Ann Arbor, MI: University of Michigan Press.

Ross, William David. 2002. *The Right and the Good*. New York: Clarendon Press.

Rousseau, Jean-Jacques. 1966. *On the Origin of Language*. Trans. John H. Moran and Alexander Gode. Chicago: University of Chicago Press.

Rushing, Janice Hocker, and Thomas S. Frentz. 1989. The Frankenstein myth in contemporary cinema. *Critical Studies in Mass Communication* 6 (1):61–80.

Sagoff, Mark. 1984. Animal liberation and environmental ethics: Bad marriage, quick divorce. *Osgoode Hall Law Journal* 22:297–307.

Sallis, John. 1987. *Spacings—Of Reason and Imagination in Texts of Kant, Fichte, Hegel*. Chicago: University of Chicago Press.

Sallis, John. 2010. Levinas and the elemental. In *Radicalizing Levinas*, ed. Peter Atterton and Matthew Calarco, 87–94. Albany, NY: State University of New York Press.

Savage-Rumbaugh, Sue, Stuart G. Shanker, and Talbot J. Taylor. 1998. *Apes, Language, and the Human Mind*. Oxford: Oxford University Press.

Scorsese, Martin. 1980. *Raging Bull*. Century City, CA: United Artists.

Schank, Roger C. 1990. What is AI anyway? In *The Foundations of Artificial Intelligence: A Sourcebook*, ed. Derek Partridge and Yorick Wilks, 3–13. Cambridge: Cambridge University Press.

Scott, G. E. 1990. *Moral Personhood: An Essay in the Philosophy of Moral Psychology*. Albany, NY: SUNY Press.

Scott, R. L. 1967. On viewing rhetoric as epistemic. *Central States Speech Journal* 18: 9–17.

Searle, John. 1980. Minds, brains, and programs. *Behavioral and Brain Sciences* 3 (3):417–457.

Searle, John. 1997. *The Mystery of Consciousness*. New York: New York Review of Books.

Searle, John. 1999. The Chinese room. In *The MIT Encyclopedia of the Cognitive Sciences*, ed. R. A. Wilson and F. Keil, 115–116. Cambridge, MA: MIT Press.

Shamoo, Adil E., and David B. Resnik. 2009. *Responsible Conduct of Research*. New York: Oxford University Press.

Shapiro, Paul. 2006. Moral agency in other animals. *Theoretical Medicine and Bioethics* 27:357–373.

Shelley, Mary. 2000. *Frankenstein, Or the Modern Prometheus*. New York: Signet Classics.

Sidgwick, Henry. 1981. *The Methods of Ethics*. Indianapolis, IN: Hackett.

Singer, Peter. 1975. *Animal Liberation: A New Ethics for Our Treatment of Animals*. New York: New York Review of Books.

Singer, Peter. 1989. All animals are equal. In *Animal Rights and Human Obligations*, ed. Tom Regan and Peter Singer, 148–162. New Jersey: Prentice-Hall.

Singer, Peter. 1999. *Practical Ethics*. Cambridge: Cambridge University Press.

Singer, Peter. 2000. *Writings on an Ethical Life*. New York: Ecco Press.

Siponen, Mikko. 2004. A pragmatic evaluation of the theory of information ethics. *Ethics and Information Technology* 6:279–290.

Sloman, Aaron. 2010. Requirements for artificial companions: It's harder than you think. In *Close Engagements with Artificial Companions: Key Social, Psychological, Ethical, and Design Issues*, ed. Yorick Wilks, 179–200. Amsterdam: John Benjamins.

Smith, Barry, Hans Albert, David Armstrong, Ruth Barcan Marcus, Keith Campbell, Richard Glauser, Rudolf Haller, Massimo Mugnai, Kevin Mulligan, Lorenzo Peña,

Willard van Orman Quine, Wolfgang Röd, Edmund Ruggaldier, Karl Schuhmann, Daniel Schulthess, Peter Simons, René Thom, Dallas Willard, and Jan Wolenski. 1992. Open letter against Derrida receiving an honorary doctorate from Cambridge University. *Times (London)* 9 (May). Reprinted in *Cambridge Review* 113 (October 1992):138–139 and Jacques Derrida, 1995. *Points . . . Interviews 1974–1994*, ed. Elisabeth Weber, 419–421. Stanford, CA: Stanford University Press.

Smith, Christian. 2010. *What Is a Person? Rethinking Humanity, Social Life, and the Moral Good from the Person Up*. Chicago: University of Chicago Press.

Solondz, Todd (dir.). 1995. *Welcome to the Dollhouse*. Sony Pictures Classic.

Sparrow, Robert. 2002. The march of the robot dogs. *Ethics and Information Technology* 4:305–318.

Sparrow, Robert. 2004. The Turing triage test. *Ethics and Information Technology* 6 (4):203–213.

Stahl, Bernd Carsten. 2004. Information, ethics, and computers: The problem of autonomous moral agents. *Minds and Machines* 14:67–83.

Stahl, Bernd Carsten. 2006. Responsible computers? A case for ascribing quasi-responsibility to computers independent of personhood or agency. *Ethics and Information Technology* 8:205–213.

Stahl, Bernd Carsten. 2008. Discourse on information ethics: The claim to universality. *Ethics and Information Technology* 10:97–108.

Star Trek: The Next Generation. 1989. "Measure of a Man." Paramount Television.

Stone, Christopher D. 1974. *Should Trees Have Standing? Toward Legal Rights for Natural Objects*. Los Altos, CA: William Kaufmann.

Stork, David G., ed. 1997. *Hal's Legacy: 2001's Computer as Dream and Reality*. Cambridge, MA: MIT Press.

Sullins, John P. 2002. The ambiguous ethical status of autonomous robots. American Association for Artificial Intelligence. Paper presented at the AAAI Fall 2005 Symposia, Arlington, VA, November 4–6. http://www.aaai.org/Papers/Symposia/Fall/2005/FS-05-06/FS05-06-019.pdf.

Sullins, John P. 2005. Ethics and artificial life: From modeling to moral agents. *Ethics and Information Technology* 7:139–148.

Sullins, John P. 2006. When is a robot a moral agent? *International Review of Information Ethics* 6 (12):23–30.

Taylor, Charles. 1985. The person. In *The Category of the Person*, ed. Michael Carrithers, Steven Collins, and Steven Lukes, 257–281. Cambridge: Cambridge University Press.

Taylor, Paul W. 1986. *Respect for Nature: A Theory of Environmental Ethics*. Princeton, NJ: Princeton University Press.

Taylor, Thomas. 1966. *A Vindication of the Rights of Brutes*. Gainesville, FL: Scholars' Facsimiles & Reprints. Originally published 1792.

Thoreau, Henry David. 1910. *Walden, Or Life in the Woods*. New York: Houghton Mifflin.

Torrance, Steve. 2004. Could we, should we, create conscious robots? *Journal of Health, Social and Environmental Issues* 4 (2):43–46.

Torrance, Steve. 2008. Ethics and consciousness in artificial agents. *AI & Society* 22:495–521.

Turing, Alan M. 1999. Computing machinery and intelligence. In *Computer Media and Communication*, ed. Paul A. Mayer, 37–58. Oxford: Oxford University Press.

Turkle, Sherry. 1995. *Life on the Screen: Identity in the Age of the Internet*. New York: Simon & Schuster.

United States of America. 2011. *1 USC Section 1*. http://uscode.house.gov/download/pls/01C1.txt.

Verbeek, Peter-Paul. 2009. Cultivating humanity: Toward a non-humanist ethics of technology. In *New Waves in Philosophy of Technology*, ed. J. K. Berg Olsen, E. Selinger, and S. Riis, 241–263. Hampshire: Palgrave Macmillan.

Velásquez, Juan D. 1998. When robots weep: Emotional memories and decision-making. In *AAAI-98 Proceedings*. Menlo Park, CA: AAAI Press.

Velik, Rosemarie. 2010. Why machines cannot feel. *Minds and Machines* 20 (1): 1–18.

Velmans, Max. 2000. *Understanding Consciousness*. London: Routledge.

Vidler, Mark. 2007. "Ray of Gob." Self-released.

Villiers de l'Isle-Adam, Auguste. 2001. *Tomorrow's Eve*. Trans. Martin Adams. Champaign, IL: University of Illinois Press.

von Feuerbach, Anselm, and Paul Johann. 1832. *Kaspar Hauser: An Account of an Individual Kept in a Dungeon, Separated from all Communication with the World, from Early Childhood to about the Age of Seventeen*. Trans. Henning Gottfried Linberg. Boston, MA: Allen & Ticknor.

Wachowski, Andy, and Larry Wachowski (dir.). 1999. *The Matrix*. Burbank, CA: Warner Home Video.

Wallach, Wendell. 2008. Implementing moral decision making faculties in computers and robots. *AI & Society* 22:463–475.

Wallach, Wendell, and Colin Allen. 2005. Android ethics: Bottom-up and top-down approaches for modeling human moral faculties. Paper presented at Android Science: A CogSci 2005 Workshop, Stresa, Italy, July 25–26. http://androidscience.com/proceedings2005/WallachCogSci2005AS.pdf.

Wallach, Wendell, and Colin Allen. 2009. *Moral Machines: Teaching Robots Right from Wrong*. Oxford: Oxford University Press.

Wallach, Wendell, Colin Allen, and Iva Smit. 2008. Machine morality: Bottom-up and top-down approaches for modeling human moral faculties. *AI & Society* 22:565–582.

Welchman, Alistair. 1997. Machinic thinking. In *Deleuze and Philosophy: The Difference Engineer*, ed. Keith Ansell-Pearson, 211–231. New York: Routledge.

Weizenbaum, Joseph. 1976. *Computer Power and Human Reason: From Judgment to Calculation*. San Francisco, CA: W. H. Freeman.

Wiener, Norbert. 1954. *The Human Use of Human Beings*. New York: Da Capo.

Wiener, Norbert. 1996. *Cybernetics: Or Control and Communication in the Animal and the Machine*. Cambridge: MIT Press.

Winner, Langdon. 1977. *Autonomous Technology: Technics-out-of-Control as a Theme in Political Thought*. Cambridge, MA: MIT Press.

Winograd, Terry. 1990. Thinking machines: Can there be? Are we? In *The Foundations of Artificial Intelligence: A Sourcebook*, ed. Derek Partridge and Yorick Wilks, 167–189. Cambridge: Cambridge University Press.

Wollstonecraft, Mary. 1996. *A Vindication of the Rights of Men*. New York: Prometheus Books.

Wollstonecraft, Mary. 2004. *A Vindication of the Rights of Woman*. New York: Penguin Classics.

Wolfe, Cary. 2003a. Introduction to *Zoontologies: The Question of the Animal*. Ed. Cary Wolfe, ix–xxiii. Minneapolis, MN: University of Minnesota Press.

Wolfe, Cary, ed. 2003b. *Zoontologies: The Question of the Animal*. Minneapolis, MN: University of Minnesota Press.

Wolfe, Cary. 2010. *What Is Posthumanism?* Minneapolis, MN: University of Minnesota Press.

Yeats, William Butler. 1922. *"The Second Coming": Later Poems*. Charleston, SC: Forgotten Books.

Žižek, Slavoj. 1997. *The Plague of the Fantasies*. New York: Verso.

Žižek, Slavoj. 2000. *The Fragile Absolute or, Why Is the Christian Legacy Worth Fighting For?* New York: Verso.

Žižek, Slavoj. 2003. *The Puppet and the Dwarf: The Perverse Core of Christianity*. Cambridge, MA: MIT Press.

Žižek, Slavoj. 2006a. *The Parallax View*. Cambridge, MA: MIT Press.

Žižek, Slavoj. 2006b. Philosophy, the "unknown knowns," and the public use of reason. *Topoi* 25 (1–2):137–142.

Žižek, Slavoj. 2008a. *For They Know Not What They Do: Enjoyment as a Political Factor*. London: Verso.

Žižek, Slavoj. 2008b. *The Sublime Object of Ideology*. New York: Verso.

Zylinska, Joanna. 2009. *Bioethics in the Age of New Media*. Cambridge, MA: MIT Press.

Index